The use of geographic remote sensing, mapping and aerial photography to aid in the recovery of blue ice surficial meteorites in Antarctica.

By the same authors

By Austin Mardon

A Conspectus of the Contribution of Herodotus to the Development of Geographical Thought 1990;
International Law and Space Rescue Systems 1991;
Kensington Stone and Other Essays 1991;
A Transient in Whirl 1991;
Alone Against the Revolution 1996;
Political Networks in Alberta 1905-1992 2002;
7 days in Moscow 2005;
The Contribution of Geography to the Recovery of Antarctic Meteorites 2005;

By Ernest Mardon Narrative Unity of the Cursor Mundi 1967;
The Founding Faculty of the University of Lethbridge 1968;
Place Names of Southern Alberta 1970;
The Conflict Between the Individual & Society in the Plays of James Bridie 1971;
Who's Who in Federal Politics from Alberta Ridings 1972;
English Studies at Canadian Universities 1972;
Community Names of Alberta 1975

By Austin Mardon & Ernest Mardon

Alberta Judicial Biographical Dictionary 1990;
Alberta Ethnic Mormon Politicians 1991;
Alberta Ethnic German Politicians 1991;
When Kitty Met the Ghost 1991;
Down & Out & On the Run in Moscow 1991;
The Girl Who Could Walk Through Walls 1991;
Alberta Mormon Politicians 1992;
Alberta General Election Returns & Subsequent Byelections 1882-1992 1993;
Edmonton Political Biography 1994;
Alberta Political Biographical Dictionary 1994;
Alberta Executive Council 1905-1990 1994;
Early Christian Saints 1997;
Later Christian Saints for Children 1997;
Many Christian Saints for Children 1997;
Childhood Memories & Legends of Christmas Past 1998;
Community Names of Alberta 1999;
Men of Dawn 1999;
United Farmers of Alberta 1999;
The Genealogy of the Mardon Family 2000;
Alberta Catholic Politicians 2000; Alberta Anglican Politicians 2001;
Liberal Politicians in Alberta 1905-1992 2002;
What's in a Name? 2002;
Edmonton Members of the Legislature 2004;
Senators and Members of the House of Commons from Edmonton 2004;
Edmonton Municipal Politicians 2005;
Alberta Francophone Politicians 2007
English Comets over the Medieval Period 2008

The use of geographic remote sensing, mapping and aerial photography to aid in the recovery of blue ice surficial meteorites in Antarctica.

Austin Mardon
Catherine Mardon

PUBLISHED BY GOLDEN METEORITE PRESS
Edmonton, Canada

A Golden Meteorite Press Book.

Cover photograph Austin Mardon, Lawrence Dommer Published by Golden Meteorite Press.

126 Kingsway Garden
Post Office Box 34181,
Edmonton, Alberta, Canada T5G 3G4 780-378-0063 aamardon@yahoo.ca

Library and Archives Canada Cataloguing in Publication

Mardon, Austin A. (Austin Albert)

The use of geographic remote sensing, mapping and aerial

photography to aid in the recovery of blue ice surficial meteorites in Antarctica / Austin Mardon, Catherine Mardon, Lawrence Dommer.

Includes bibliographical references. ISBN 978-1-897472-15-3

1. Meteorites--Antarctica. 2. Remote sensing--Antarctica. 3. Aerial photography in geology--Antarctica. 4. Geological mapping--Antarctica. I. Mardon, Catherine A., 1962-

II. Dommer, Lawrence III. Title.

QB755.5.A6M37 2011 551.3'97 C2011-905084-6

Dedicated to:

Ernest George Mardon
my
first & best professor.

Acknowledgements:

Thanks to my fellow members of the 1986 ANSMET Expedition to the Lewis Cliff, Antarctica.

Forward:

This book comes out of my youthful membership on the ANSMET 1986 Antarctic Meteorite Recovery Expedition to the Lewis Cliff Ice Tongue. The historic importance of the Antarctic meteorite recovery program will be acknowledged in the annals of historic planetary geology. It has gone from two countries engaging in recovery to several countries having concerted efforts at recovering meteorites. Much of what has been written is culled from personal research over the years.

This paper describes the contribution of geography through the use of cartography, remote sensing and aerial photography to the ongoing recovery of meteorites in the Antarctic. It gives details of the first geographer to go on an Antarctic meteorite recovery expedition experiment relating to the aerial detection of meteorites from aerial photography and the ground truth experiment comparing meteorites and other surficial terrestrial rocks. While that experiment was null it did point to using criteria to select sites for the recovery of meteorite placers along the Trans-Antarctic mountain range. Meteorites have been estimated in the Antarctic to amount to over 100,000 potentially recoverable ones. Over the last thirty years their have been over 25,000 meteorites recovered. The specific scholarly contributions by the author, amounting to twenty three peer reviewed scholarly communications, to the study of meteorite recovery are discussed in length.

Austin Mardon, C.M.

Table of Contents:

Ice Moved
Remelting of blue ice areas
Cryoconite holes
Aeolian moved
Direct falls onto ice
Katabatic Winds
Scouring
Melting
Sublimation
Transportation to and from placer sites
Using C-130 for transport
Using Twin Otter for transport
Long Range surface traverses
Robotic Contribution to Antarctic Meteorite Recovery Program
Robot surface traverses
Robotic aircraft for surveying
Analogue Terrestrial comparisons with Antarctic Meteorite Recovery locales
United States expedition to the Greenland Ice cap to recover Meteorites
Pleistocene deposits of meteorites in Canada
Columbia ice field in Canada
Recovery of meteorites in Canadian Arctic
Pleistocene deposits of meteorites in Russia
Meteorites lack terrestrial contamination
How meteorites are archived and used after recovery.
Helicopters being used to find meteorites
Side mounted video cameras
Persons flying
Aerial Photography Contribution
Two large meteorites might show up on regular aerial photography
Archived aerial photography
Current aerial photography
Remote Sensing Contribution
Satellite imaging contribution

Problem:

The problem as discussed in how to describe the contribution of geography through mapping, remote sensing and aerial photography to the international Antarctic meteorite recovery effort. This problem has and will contribute directly to the international recovery of precious meteorites found on the surface of blue ice placers. The meteorites are found in an extremely dangerous and treacherous region of the Earth the Antarctic polar plateau. A major question is how geographic techniques can be applied to the study of recovery methods and the understanding through physical geography of the dynamics of Antarctic meteorite blue ice placers.

The research spans over seventeen years from 1986 through to 2004. Many of the published scholarly communications were published through meetings at the Lunar and Planetary Institute, Houston, Texas, United States of America or published at conferences partially sponsored by the Lunar and Planetary Institute. The author hopes to show the over twenty scholarly communications show how geography has contributed to the recovery and understanding of Antarctic meteorite processes.

Introduction:

The largest source of extraterrestrial materials has come from the Antarctic over the last thirty odd years as meteorites. (Mardon, 1996) Our only source of rock samples from Mars comes from the over ten thousand meteorites that have come from the Antarctic meteorite surface placers. The pristine condition of Antarctic meteorites and the varied nature of the sources have meant that the United States of America Antarctic meteorite recovery program is a scientific priority. The first Antarctic meteorite was recovered in the early 20th century by the famous Australian Antarctic explorer Douglas Mawson's expedition. Meteorites have been estimated in the Antarctic to amount to over one-hundred thousand potentially recoverable ones over the last 30 years and the next twenty years. Antarctic meteorites have become a major source extra-terrestrial samples from Mars, the Moon, from Near-Earth objects and from the Asteroid belt. (Mardon, 1992b)

Recovery of Blue Ice Meteorites in Antarctica

Geography can and has contributed to the recovery of meteorites in the Antarctic thus benefiting the sum of mankind's knowledge planetary and other body geology of the inner Solar system. According to Harvey (2003), "Antarctic meteorites rival the Aollo lunar specimens in their impact on planetary materials research while new speciments continue to arrive yearly." One sample of a meteorite from Mars gave the first physical evidence for the possibility of previous life on Mars.

History of research:

The author in 1986 was the first geographer to go on an Antarctic meteorite recovery expedition. (Mardon, 1987d) He was a field member for just under two months at the Lewis Cliff Ice tongue meteorite placer site. The United States Antarctic Research Program (USARP) team recovered over 700 meteorites in that locale over the season of 1986-1987. There was an aerial over flight where photographs of the Lewis Cliff ice tongue were taken. In the field during the season their was ground truth images taken.(Mardon, 1987f) The first aerial photographs that were taken of the Lewis Cliff ice tongue that were taken were in the late 1940's related to Operation High jump and American expedition. Those images and successive images are archived at the United States Geological Service(USGS) in Reston, Virginia, USA. The purpose of the research was to see if it would be possible to detect meteorites from the air so that later they could be recovered knowing exactly where they were.(Cassidy, 1987; Mardon, 1987b) Sadly this experiment was unsuccessful. But in other areas such as looking for sites where meteorites can be recovered from geography through its disciplines of cartography, aerial photography and remote sensing. Certain areas of the Antarctic seem to concentrate meteorites and they are also exposed at these locales. Geographic tools can be used to select these sites for field reconnaissance.

The blue ice surfaces that meteorites are found on are not static surfaces but meteorites move disappear from the surface from melting into 'Cryoconite' holes. These phenomena are when the meteorites melt up to one foot down into the ice and are then covered with ice that re-melts over the sunken meteorite. Meteorites then are re-exposed. (Mardon, 1998b) This phenomena also means that placers that are completely surveyed for meteorites might in the future even the next year. Also meteorites are transported by the ice surface and also theoretically could be moved by the severe katabatic winds. Meteorite fall fields in Australia have had the occurrence of meteorites being moved along the surface of the Australian desert by wind. (Mardon, 1991) The ice surface of the meteorite placers is sublimated, melted and scoured by the katabatic winds. The meteorites that appear are not found where they fall but are concentrated into upwelling ice areas near mountain ranges by the slow moving of the ice. An analogy of what occurs is that the vast Antarctic polar plateau is an ocean and the meteorites fall on the surface and then get concentrated on the edge of the Antarctic polar plateau by the slow moving movement of the ice cap in the same way that driftwood is concentrated on the shore of the ocean. The meteorites stay in the surface zone of the ice cap that is eroded away and flows toward the edge of the Trans Antarctic mountain range. At these zones the ice is gradually eroded by the katabatic winds.

Purpose:

At the first stage of selection of a potential field site to look for meteorites the researcher looks for the characteristics of an Antarctic meteorite placer area. They have over-flights, look at maps and examine aerial photographs and other remote sensing images. (Mardon, 1987a) The majority of meteorites that have been recovered by the United States Antarctic meteorite recovery program are along the Trans-Antarctic Mountains. The meteorites are concentrated along the edge of this mountain range and are exposed by the ablating process of katabatic winds along this mountain range.(Mardon, 1991) The criteria of selection along this range of blue ice meteorite placers is fairly consistent. Once meteorites are found, maps of where they are exactly found are mapped out and maps of these meteorite placers are made. New sites are selected based on the criteria that has been deduced from previous locations where meteorites have been found is made.

These abstracts and articles that were presented were the field report on the season that the author was a member of. Over seven hundred meteorites were recovered during that field season at the Lewis Cliff Ice tongue meteorite placer. It was published less then six months after the January end of the field season.(Cassidy, 1987; Mardon, 1987b) there were six members of the team. One team member was from Japan, one was from Austria, one was from the Netherlands, two were from the United States of America, and the author who was from Canada. The recovery of Antarctic meteorites has given Inner solar system geology a good picture. The most noteworthy recovery is that of SNC meteorites which have been shown to be from Mars as a parent body. A large body would have impacted Mars and Mars surface material would have been ejected from the surface of Mars due to the force of the impact and then the rocks would eventually leave Mars orbit and hit the Earth.(Cassidy, 1987; Mardon, 1987b)

Cartography contributes to the recovery using maps to discover meteorites and also contributes by mapping out the locations of where the meteorites are found in the field. Maps are made of the meteorites and where they are found. By doing this it is hoped that scholars will get a better picture of the mechanisms of how meteorites are deposited and concentrated. This in turn will lead to better site selection criteria.

Methods:

The author will examine the techniques that have been taken from the discipline of geography and examine and contrast how they have been used to enhance the recovery of meteorites from the Antarctic. Ground truth was taken in the field and mapping techniques were done. There was examination of geographic techniques. Field mapping was done of the Lewis Cliff ice tongue during the season of December 1986 through January 1987.(Mardon, 1988g)

The usual method of looking for meteorites is divided into three phases.

The first is when potential blue ice meteorite placers are looked at by looking at maps, aerial photographs, satellite imaging and discussions with previous expedition members who have visited the potential recovery search areas. The human input is useful in that human exposure might give an idea whether the potential blue ice areas seem to have the characteristics that might indicate that there are meteorites on the blue ice areas.

The second phase is where there is a team that is sent out to the field where the potential blue ice meteorites might be found. This is a reconnaissance mission where surfaces are crossed in some cases several sites are visited and the team has mobile characteristics. The camp might be moved several times. Extensive methodical searchers are not conducted but sampling traverses of the potential sites occur. The teams are moved around now by fixed wing aircraft and extensive traverses occur when a reconnaissance occurs. If meteorites are found then in subsequent seasons more concentrated examination of the potential placers occurs.

The last and final phase is where extensive, systematic, comprehensive and concentrated surveys of areas that provided recovered meteorite samples previously in reconnaissance traverses occurred. This comprehensive surveying and meteorite recovery was the majority of the time and effort of the season that I was a member of in the 1986 – 1987 austral summer. We did make several side trips one of which was to the Goodwin Nunatak as reconnaissance traverses.

It is also possible that meteorites keep getting exposed in the blue ice placers due to sublimation and melting of the blue ice areas. There was a proposal two years ago to have a private expedition to the Patrick Hills meteorite complex which had already been surveyed and meteorites recovered from the surface. The idea was that new meteorites might be exposed from underneath snow pack and exposed due to sublimation and melting of the blue ice.

The author has lobbied for the last sixteen years in Canada that Canada should have an independent Antarctic scientific research program. Meteorite recovery could be one of the possible programs if Canada was to enter into a cooperative program with another nations program. The American Antarctic program relies on Canadian aircraft in the Antarctic for civilian flights including recent rescue flight to the South Pole during the southern regions winter. Canada has a robust Arctic research program and could bring to the

Antarctic international research table a wealth of expertise in the area of polar research.(Mardon, 1995)

The United States Antarctic meteorite recovery program has historically included a substantial international flavor. In addition to this at least half of the field members would be experienced polar and specifically Antarctic experienced veterans.(Mardon, 1996) Science erroneously reported that the Americans were the pioneers of first finding meteorites in the Antarctic. It is in reality the Japanese Antarctic program that first engaged in systematic recovery efforts to find meteorites. They also cooperated substantially with the American Antarctic meteorite recovery program in the beginning under Dr. William A. Cassidy. Also of course as previously stated the Australian Antarctic explorer early in the 20th century accidentally came across a meteorite.(Mardon, 1996)

The following section details the results of the research related to the geographic contribution of recovery of Antarctic meteorites.

Results:

The primary result is that thousands of meteorites are being recovered from the Antarctic. Cartography is used to find locales that meteorite placers occur at. Maps are made of meteorite placers and the meteorites that are found there are located on these maps. Remote sensing systems are used to find the blue ice placers where meteorites that fall all over the interior of Antarctic are found.(Mardon, 1988e) The size of meteorites that are recovered can not be distinguished from photographs taken from airplanes at heights averaging 500 feet.(Mardon, 1988b) Criteria of selection of areas that receive ground reconnaissance use cartography, aerial photography and remote sensing. (Mardon, 1987e) The author has proposed the increase in the number of countries Antarctic programs that hunt for meteorites to include the French and Russian Antarctic programs. (Mardon, 1992a; Mardon, 1995; Mardon, 1998a; Mardon, 2003) Patterns of recovery can be replicated in other areas of the world such as North America or Russia. (Mardon, 1997) Previously glaciated regions of the world might have meteorites that might be able to be recovered. (Mardon, 1988c; Mardon, 1988d) The geology of the placer deposits can be known to a greater extent such as the process of deposition of meteorites in Antarctic placers. (Mardon, 1998b)

The author in 1985 proposed to the team leader of the United States of America's Antarctica meteorite recovery program that limited targeted aerial photography is used to directly detect blue ice surface meteorites. In many cases meteorites recovered on the surface of the ice have black fusion crusts still intact. In many cases they are the only rocks on the surface of the ice. It was proposed that either fixed wing aircraft such as the C-130 Hercules aircraft or rotary wing aircraft-helicopters be used as platforms for over-flights of blue ice areas.(Mardon, 1987a)

Current Antarctic Nations involved in Recovery of Antarctic Meteorites:

Up to the time when I went to the Antarctic in 1986- 1987 there had been only two nations that had mounted dedicated Antarctic meteorite recovery expeditions. Those two nations were Japan and the United States of America. Since then their have been other countries and little known private meteorite recovery ventures that have recovered meteorites in the Antarctica. There has been the EUROMET program from Europe, a dedicated Italian program, recently China recovered a Martian meteorite in Antarctica and some little known private expeditions to the Antarctic. It is important that the different Antarctic meteorite recovery programs have similar recovery protocols and that meteorites be open to scientists from all over the globe. It is possible that more nations will have active meteorite recovery programs and as it has been with current missions to the Antarctic to recover meteorites international cooperation among meteorite recovery scientists occurs a lot. The Italian program cooperated early on with the EUROMET Antarctic meteorite recovery program.

The following nations have substantial Antarctic meteorite recovery programs.

Japanese Antarctic Meteorite Recovery Program:

The Japanese Antarctic meteorite recovery program has been active since the early 1970's in Antarctica.(Kojima, 2000)

They have focused on sites near and around the Yamato Mountains. The Japanese program was the first dedicated Antarctic meteorite recovery program in the Antarctic. The logistical structure of the Japanese program is different then other programs in that they do surface traverses to the meteorite blue ice standing surfaces and they get transported to the Antarctic by ship. It involves a much larger commitment of time then the two to three month commitment that the American program entails in transit and field time in the Antarctic. Also they do not have as much flexibility in terms of sampling geographically separated sites because they do not use aircraft but rely on surface traverses. They also do not catalogue each individual recovered sample as exactingly as the American program does. The Americans use the lunar sample processing rooms at the Johnson Space Centre for first and primary cataloging of all samples. The Japanese program focuses most of their geochemical efforts on apparently distinct and unique samples rather then looking at the entire set of samples of meteorites that have been found in the Antarctic. While it is scientifically important to recover unique samples and focus the bulk of post recovery scientific analysis on unique geological samples it is important also to at least do some primary processing of all samples especially because it might be that some unique samples and scientific patterns might not be immediately apparent on 'first blush.'

The Japanese program does not have the same standard of sterile collection as the American program. This is primarily due to the fact of their logistical limitations.

It should be noted that it was the Japanese Antarctic geologists that first recovered the meteorites in any number in the Antarctic and had the 'Eureka' moment when there was a link between blue ice and substantial surficial deposits of meteorites on blue ice fields. They galvanized the American researcher Dr. 'Bill' Cassidy with the idea of a potential American effort to recover meteorites in the Antarctic. They jointly worked with the Americans to recover meteorites in the early years. I spent my field season, in a tent, in Antarctica at the Lewis Cliff Ice Tongue with the legendary Japanese geologist Dr. Keizo Yanai. The Japanese and Americans have cooperated with each other since the early 1970's. The Japanese have focused their efforts in expeditions that were near their base on the other side of the Antarctic across from the main belt of other searched sites on the Trans-Antarctic mountain range. The Japanese program does not go down each year but goes down to the Antarctic every couple of years.

The Japanese model of recovery of meteorites using surface traverses to arrive at meteorite blue ice standing surfaces is a model that other countries could emulate in their potential meteorite recovery programs in the Antarctic. It is cost effective in that it does not use expensive aircraft and can be used by any nation with surface traverse capability. The Japanese were the first to conduct large scale recoveries of meteorites in the Antarctic. They are the pioneers in the field of Antarctic meteorite recovery.

Without Japanese tenacity and originality today it is unlikely that we would have the scope of recovery of the many tens of thousands of separate non-terrestrial samples that have been recovered from the Antarctic.

United States of America Antarctic Meteorite Recovery Program:

Since the 1970's when the Americans first went down to the Antarctic with a dedicated Antarctic meteorite recovery expedition their has been a meteorite recovery program every year except for one when many United States of America Antarctic programs were canceled due to a season that was affected by the Sun. Every year their since then their has been an expedition.(Cassidy, 2003)

The American program is the 'gold standard' of Antarctic meteorite recovery in the Antarctic at all stages. They use small and large aircraft to travel to meteorite stranding surfaces all across the Trans Antarctic Mountain range. They sometimes split their team and go to different blue ice placers. Usually they have a team doing reconnaissance on one or more potential meteorite placers and then they have the remaining portion of that season meteorite recovery team doing a systematic search of a known blue ice area. The American program in contrast with the, especially, Japanese program has a substantial foreign presence. This is partially due to the fact that the Japanese have to go down to the Antarctic for under two years as compared to the Americans who at the most only have to go down for three to four months.

The American program has acquired a substantial portion of the Antarctic meteorites over the year. Several unique meteorite samples have been recovered including several Lunar samples, several Martian samples, and several very rare meteorite samples that are in unique categories of classification.

The Americans are the leaders currently of Antarctic meteorite recovery. They have examined and discovered disparate meteorite blue ice standing surfaces. They have mapped out in great detail in the field where meteorites are found. This has resulted in a greater understanding of how meteorites are concentrated and exposed. Within the last several years more complex models of how meteorites are concentrated have been explored and discussed in print. The next advance is that each sample is examined and cataloged at the Johnson Space Centre. Each sample is classified and results published by the same laboratory that curates the Lunar samples that were brought back from the Moon by the Apollo missions. Then samples of meteorites recovered are released to the scientific community. Any researcher from any country in the world can apply through the Meteorite Working Group, headquartered at the Johnson Space Centre, to acquire a sample or samples of an Antarctic meteorite to conduct research. By examining and classifying every sample it is possible to discover the characteristics of the entire sample of all meteorites that have been recovered from a meteorite stranding surface. Other countries that have Antarctic meteorite recovery programs do not examine and classify each sample. They focus on the unique samples and leave the other samples unexamined. Scholars will only understand the characteristics of falls over the last two million years which encompasses all of the proposed recovered meteorites by classifying and describing each sample. Also some samples might appear as a standard meteorite whose classification has appeared frequently in the Antarctic meteorite collection before but with detailed classification it might show some unique or scientifically interesting characteristic. After research interest of a meteorite at the Johnson Space Centre has waned then the samples are sent to the Smithsonian Institute for currating.

The American thoroughness in mapping out meteorite locations and detailed geological classification of each recovered meteorite sample has led to a greater understanding of Antarctic meteorite stranding blue ice surfaces. Our current understanding of meteorite stranding surfaces in detail comes from the detailed mapping by the American program in the Antarctic.

Chinese Antarctic Meteorite Recovery Program:

The Chinese Antarctic meteorite recovery program has had little published on it. Recently according to Satterwhite (2004) a member of the meteorite working group at the Johnson Space Centre in Houston the Chinese recovered a Martian meteorite in the

Antarctic. Aside from this little is known about the Chinese Antarctic Meteorite Recovery Program in the Antarctic.(Russell, 2002) The author six years ago talked to the deputy director in Peking by telephone about the possibility of a Chinese Antarctic program.(Mardon, 1998) In 1999 and 2000 had expeditions to the Grove Mountains region. (Harvey, 2003, 96)

Italian Antarctic Meteorite Recovery Program:

It should just be noted that Italy has had several Antarctic expeditions that have recovered meteorites from the Antarctic.(Folco, 2002) The Italians have also participated in the EUROMET Antarctic meteorite recovery program. The Italians participated with the EUROMET and in addition had independent searches in 1997, 1999 and in 2001.

EUROMET-European Antarctic Meteorite Recovery Program:

A consortium of European countries have gathered together and had several expeditions to the Antarctic to recover meteorites. (Folco, 2002) They searched several time in the 1988-1989 field season and then in the early 1990's.(Harvey, 2003, 96)

Proposed other National Future Antarctic Meteorite Programs:

With the use of smaller less expensive aircraft such as the Twin Otter it has become possible to transport small Antarctic expeditions. This, be it, whether they be private, non-governmental supported expeditions or other non-represented countries that could mount their own expeditions. Since the author went to the Antarctic with the Americans to recover meteorites several other countries have started Antarctic meteorite recovery programs. It is currently an opportunity for meteorites to be recovered to a greater extent then they have been. While the American National Science Foundation program is the 'gold standard' as far as how the recovery should happen, collection, mapping of standing surface meteorite occurrences and processing of all samples; this does not happen with other programs. Their still should happen that recovery should be achieved by more countries. Antarctic Meteorites are a "Poor Man's Space Program," and as such they should be looked at as a potential bonanza scientifically for scientists from all over the world. More nations should have Antarctic Meteorite recovery programs!

Private American Antarctic Meteorite Recovery Programs:

I have heard of private expeditions that have gone to meteorite standing surfaces or placers.(Moser, 2002) But there is no literature on what they recovered or whether sadly whether the samples that they recovered were used for scientific purposes or whether they were just put into private hobby collection. The American program has made a longstanding policy of distributing Antarctic samples to the broadest segment of international scientists to study the samples. The samples that are acquired and recovered privately do not necessarily get studied. There are many examples of this from meteorites that are recovered privately not being given to scientists for studied. Also, the market for these samples are such that they dwarf the budgets of scientific researchers. Unique geological samples should be the common heritage of mankind and the scientific use of these meteorite samples should be more important then turning samples into slides which then lie on someone wall or bookcase or desk as a odd curio. Dr Cassidy (2003) says that there should be the opportunity for private individuals to go to the Antarctic with national meteorite expeditions as long as they fill the general requirements. Dr. Cassidy(2003) is concerned that the privately recovered meteorites will not be curated as thoroughly as they are presently done by governmental programs. The cost to join a private expedition to recover meteorites in the Antarctic is about $50,000 American.

Russian Antarctic Meteorite Recovery Program:

The Russian Antarctic program has had a substantial Antarctic presence dating back to the former Soviet Union. They became active in the Antarctic in the modern period after the International Geophysical year in the late 1950's. They have a station in the interior called Vostok Station where they have surface traverse to each year to re-supply it. The Russian Antarctic program has the logistical, personal experience and scientific experience to mount a successful Antarctic Meteorite recovery expedition. EUROMET the European Antarctic Meteorite recovery program could be a good jumping off point. If cost was a factor in the equation as to why they have not set up a program then they could sell selected slices of recovered meteorites on the private market and use the remaining samples for research purposes. Tunguska was the most notable historical impact of a bolide on the Earth during recorded history and it happened in Russia's back yard in Siberia.

The potential requirements of the logistics would either be transit to potential blue ice surface placers by air or by surface traverse. The Japanese Antarctic program uses surface traverses to the Yomata Mountains meteorite standing surface. They get to the Antarctic by ship and then traverse to the interior where meteorites are found. This is an option that the Russian Antarctic program could implement without any additional major logistical expense. They could also use small aircraft injections into blue ice meteorite placers but this would likely cost more money. Russia is a major Antarctic researcher and it is a gap that they do not have an Antarctic meteorite recovery program. Blue ice standing surfaces occur all along the Trans Antarctic Mountain Range and rather then as some would view it is a competition it should be viewed as getting the samples faster then they would be otherwise recovered if at all. It has been shown that ski-equipped aircraft can land on blue ice areas and snow packed snowfields even with substantial sastrugi.

A Russian Antarctic meteorite recovery program could also pay for itself by incorporating private individuals willing to pay the cost of mounting such an expedition. Eco-tourism and Antarctic tourism has proven that people are willing to pay substantial money to go to the Antarctic. I would estimate that an expedition could charge upwards of 50,000 dollars per person. This would likely pay for the cost of the expedition in addition to, as I previously said, selling a portion of the meteorites that are found.

Currently Lunar samples that were brought back from the Moon by Soviet robotic space vehicles are archived in Moscow. The Russian academic community does have the interest and capacity to examine and utilize any meteorites that would be recovered. It would also open up the chance to cooperate with other nations geochemists by being able on an equal footing to exchange samples and give as well as take in the area of geochemical research.

In this era of cost cutting of research especially in Russia and the C. I. S. an Antarctic Meteorite Recovery Program could quite feasibly be a cost neutral endeavor if some creativity was brought to bear on it being set up.

French Antarctic Meteorite Recovery Program:

The French government has had a longstanding robust Antarctic scientific program. They have had French scientists participate in both the United States of America Antarctic meteorite program and the EUROMET meteorite recovery program. The author believes that the French Antarctic program should try to implement an independent Antarctic meteorite recovery program.(Mardon, 2003) The likely type of French logistical support for a field party would be a surface traverse or the use of small aircraft such as the Twin Otter from Ken Borak Air based out of Calgary. The author does not know whether there are blue ice areas near the major French Antarctic station on the edge of the East

Antarctic Ice Cap. It is likely that there are similar deposits that are at 2000meter altitude with under ice barriers that would result in Antarctic stranding blue ice meteorite placers. The author if the French government were to set up an independent Antarctic meteorite recovery program to use Twin Otter aircraft as several blue ice sites could be visited in one season around their station and along the edge of the Trans-Antarctic Mountain Range. The Italian Antarctic program has had an Antarctic meteorite recovery program and the French program should join those nations that have stepped up to the plate and has supported an independent Antarctic meteorite recovery program.

French logistical experience in the Antarctic could support an independent robust Antarctic meteorite recovery program.

Ukrainian Antarctic Meteorite Recovery Program:

An independent Ukrainian Antarctic meteorite recovery program could at a cost neutral level be set up by making arrangements for private individuals to pay up to $50,000 to be a member of a governmental Antarctic meteorite recovery expedition. Also a portion of the meteorites could be sold on the private market after the expedition returned from the Antarctic recouping and covering any costs incurred by the Ukrainian government in mounting an Antarctic Meteorite recovery Expedition. The logistics of getting to the Antarctic could be that they would have to fly into a base in the Antarctic peninsula and then fly into the interior mountain range, the Trans-Antarctic mountain range. There are blue ice meteorite stranding surfaces on the side of the East Antarctic Ice sheet near to the Antarctic peninsula. A flight in for five to six expedition members is the only way for an expedition because the Ukraine does not have a major base in the Antarctic. The value of a Martian meteorite would likely be more then one million dollars on the open market with some of the meteorite being kept for scientific purposes. Because the Ukraine does not have a substantial logistical and scientific presence in the Antarctic it means that they would have to have a stand alone expedition searching for meteorites except for flying privileges into a base tentatively in the peninsula on the way to flying into the interior ice fields such as the Patrick hills which could start in South America then flying south. Many blue ice areas have had reconnaissance on them for meteorites but have not been searched in detailed full search by a grid search pattern.

The Ukraine government could with some creative financing and sponsorship mount an independent Antarctic meteorite expedition to one or more of the Trans Antarctic Mountain range meteorite stranding blue ice surfaces.

How Antarctic Meteorites are found:

There was no indication in Antarctica until the Japanese started to recover large numbers of meteorites that they existed in such high density in blue ice stranding surfaces in the Antarctica. In all humility there should have been previous to the 1970's realization that they existed in such abundance in the Antarctic. The Lewis Cliff Ice tongue is just to the East of where Scott ascending the Beardmore Glacier on his ill-fated expedition to the Antarctic South Pole. Many of the areas where they were found had been traversed on the surface if not once but several times by previous expeditions. It was not until the fortuitous discover of the Yamota meteorites that the gaze of meteorite specialists started towards areas of the Antarctic near stations that had blue ice areas. Simply meteorites are found on the ground by searching each rock on blue ice or moraines sometimes on ones hands and knees examining the specimens. In some cases this process is made easier as the only rocks on the blue ice stranding placer are meteorites. Many new sites have been found with serendipity when other researchers find a single meteorite and then a reconnaissance of the stranding surface occurs and then if some additional meteorites have been found then a systematic search occurs. The author proposed the unsuccessful concept that meteorites could be found by using aerial photography. Aerial photography had been used to find the actual meteorite stranding surfaces.

NOMAD has been tested and it is possible to find meteorites using a robot. The main problem is that it takes as many people to keep the robot going as it takes to main a human conducted search.

<u>Through long range surface traverses and reconnaissance</u>:

Recovery of Blue Ice Meteorites in Antarctica

The first stage in finding meteorites in the field is to mount a reconnaissance of an area. The area is walked over or driven over in a cursory manner by snowmobile and/or by foot. The driving over is not systematic. In some cases large numbers of meteorites are found just by randomly crossing a potential meteorite standing surface. Some meteorite placers only rocks on the surface are meteorites. This depends on whether the upward ice flow to the stranding surface has any terrestrial outcropping that might be gardened in with the meteorites.

Today all field team Antarctic expeditions are acquainted with the potential that the areas that they cover might have meteorites on them. The presence of meteorites in Antarctic has gotten a lot of press in the media. Several new sites have been discovered by researchers accidentally finding a meteorite and then either notifying an existing Antarctic meteorite recovery expedition. Meteorites generally are found on blue ice stranding surfaces and in the earlier period of Antarctic exploration these areas were avoided because dog teams found it difficult to traverse also these areas have associated covered and open crevasses near or in them.

The two methods of getting to an blue ice surface is by flying in or driving a snowmobile. For sites near to each other within ten to twenty miles it is better to drive for two reasons first is the cost and the second is that during the snowmobile drive the party that is making the traverse can be looking for any meteorites. While I was in the Antarctic we found an approximately one kilogram meteorite lying by itself near the Goodwin Nunatak on the other side of the valley from the Lewis Cliff ice Tongue.

Through detailed full search of small area:

In the year before I went down to the Antarctic major portions of the upper Lewis Cliff ice tongue were systematically searched and yet even with that when going across the ice tongue each day we found several meteorites that had either been exposed during that year or had been missed by the systematic search. When a promising site for meteorite recovery is identified a later systematic search occurs. The method of such described locales is as follows, "Field team members form a line, each member a few to several tens of meteors apart depending on the number distribution of rocks to be examines and the method of travel (on foot or motorized vehicle). The team then proceeds across the meteorite stranding surface in a direction perpendicular to this line examining each rock in their path. After each transect… the team shifts over to cover a new parallel transect."(Harvey, 2003, 96) This is called the transect sample model that is used in other geological searches of a surface.(Harvey, 2003) This transect sample model is the traditional method of examining a surface for geological samples.

Even with detailed searches of a Antarctic stranding blue ice surface it is possible to miss areas that are covered with some temporary snow drifts such a sastrugi. In later years these overlay of snow might be either melted away or blown away.(Harvey, 2003) This means that it might be possible that subsequent searches of blue ice areas might reveal new meteorite samples in some major concentrations. Their was one year under Dr. Cassidy's (2003) leadership that the reconnaissance team found more meteorites then the systematic search team that were searching different stranding meteorite blue ice surfaces. This was due to the different weather at the two sites and a very successful reconnaissance search. Because at the same stranding surfaces there is the very good likelihood that there are different meteorite falls it is likely in the future that previously searched sites might merit being searched again and possibly again for meteorites. Not all meteorites found at specific stranding surfaces are from the same fall so it is very likely new meteorite falls might be found. In other fall recoveries in non-Antarctic locales it is likely that meteorites found near each other are from the same parent body and broke up during the entry through the atmosphere. In summation if a site that is easy to get to logistically it might merit a second or third systematic search with a secondary team to see if it is possible to find more meteorites. Sites that this might qualify for would be sites within helicopter range of the American station at McMurdo Station on Ross Island, Antarctica. It could be said that in the future if the demand for meteorites continues then the above might happen in detail where meteorite search teams might search previously intensively searched areas. When I talked to one of the private search team leaders I found out that they do not go to new sites or conduct reconnaissance they go to sites that have either been entirely intensively searched or partially intensively searched.

On foot:

It is sometimes necessary to be on foot transecting a meteorite stranding surface. A situation such as this is when logistically it is not possible to use a snowmobile when for example a helicopter took a team in and then left the team. They would then have to search firn and blue ice areas for meteorites by foot. Another situation that I encountered in the Antarctic is when we found meteorites some of which were very small in a moraine, later called Meteorite Moraine. This placer was to the west near to the Lewis Cliff Ice Tongue. It was not possible to drive over the moraine edge as their were to many rocks and the meteorites were very dense in concentration, small in size and a person had to search the moraine on ones hands and knees to distinguish between terrestrial and non-terrestrial rocks. The searching on foot is still done by the transect sample model back and forth. Newly exposed meteorites and meteorites that have somehow been missed by the first search might mean that if a field team has time they can go back to an area especially a location such as Meteorite Moraine with such concentrated samples. It is very difficult to search a moraine by foot because each rock has to be looked at while on foot. The author believes that moraines likely contain substantial surface deposits of meteoritical material but it is to difficult to distinguish between non-terrestrial and terrestrial rocks. Something like a moraine is a suitable place for the development and deployment of robotic recovery systems such as NOMAD. Being on foot on ice is in itself potentially harmful as it is possible to fall due to the slippery surface. Also it is more possible to fall into a snow covered crevasse if one crosses by foot due to the increased weight for imprint size. I nearly fell to my death early on when I got off of my snowmobile and accidentally stood on a covered crevasse. A snowmobile spreads out the weight of a person and the vehicle so the pound per square inch pressure on a snow-bridge is less when one is on a snowmobile as compared with foot transportation. It is much safer to stay on a snowmobile as compared with foot on the heavily crevassed blue ice meteorite stranding surfaces.

On snowmobile:

The snowmobile with towed Nansen sleds is the form of transportation around locales in Antarctica after aircraft or surface traverse to a potential meteorite stranding surface. I logged over 1,100 kilometers on my Canadian built snowmobile or as they are known in Canada, Ski-Doo. The snowmobiles were Bombardier custom snowmobiles. They had some unique custom made changes to them. In the treds they had bolts screwed and stuck out so that the snowmobiles would hopefully get more traction on the blue ice areas and on firmly packed snow drifts- sastrugi. The usual method of travel on a snowmobile was to kneel with one leg on the seat of the snowmobile and the other leg all the way down to the running board. The reason for this was two fold; the first reason being that in case the snowmobile was to crash through a snow-bridge over a crevasse the theory is that the person could jump from the snowmobile easier then if they were sitting astride the seat of the snowmobiles and the second reason being that a person is a bit higher by kneeling on the seat and it is easier to see meteorites and avoid any of the signs of hidden crevasse fields. The snowmobile is the workhorse of the meteorite recovery program.

The snowmobile is also safer to be on a snowmobile then on ones feet on crevasse field because the pounds per square inch pressure is less per square inch for a snowmobile then the pounds per square inch pressure for someone that is standing on their feet. The author had a crevasse collapse underneath me when I stood next to my snowmobile and the bridge did not collapse under my snowmobile until after my snowmobile got to the lip of the hidden crevasse snow-bridge. This incident happened on the third day of the field season when we were making a traverse on the southern portion of the Beardmore Glacier from where the C-130 Hercules put us in and our season field camp near the Lewis Cliff Ice Tongue. I was very lucky that I did not perish. The mountaineer expert later said that I was very lucky to have not been swallowed up by the crevasse that day. That was only the beginning of a very challenging season in the Antarctic at the Lewis Cliff Ice Tongue.

Description of day to day in Antarctic searching for meteorites:

Recovery of Blue Ice Meteorites in Antarctica

My experience in the Antarctic under Dr. 'Bill' Cassidy was one of the most mentally and physically challenging experiences of my life. We found over 700 meteorites during the field season. Several days we worked around the clock to search the areas by snowmobile because when the wind was not blowing was a time when it was not as cold because the wind chill was not as bad as when we would get winds of up to 100 kilometers per hour. We worked in the area around the Lewis Cliff Ice Tongue Complex. My odometer logged me at having traveled over 1,100 kilometers that season around that area. When a meteorite was found unless it was on a reconnaissance traverse a flag pole was put sticking up out of the ice next to the meteorite. Then later the meteorite would be collected using surplus Apollo Moon mission clean bags with minimal contact and its exact location was mapped from one of several established benchmarks. Maps would be created from this data that would be eventually used to develop a better understanding of Antarctic meteorite stranding surfaces.

When the weather was nice these opportunities for good collection times would not be wasted. A person was never left alone in the field. Anytime someone left the camp there would be at least two people with snowmobiles in case one snowmobile broke down.

Trip to the Antarctic continent and the interior by aircraft:

The flight to the Antarctic in ski equipped C-130 Hercules left from Christchurch, New Zealand. I had flown there from the continental United States of America. The flight from Christchurch eventually arrived at the ice shelf runway at McMurdo Station, Ross Island, Antarctica. Just on the other side of the isthmus that McMurdo is on the New Zealand government has a station. They share logistical infrastructure with the Americans and try to compliment scientific research and logistics at their stations. A dedicated over flight over the Lewis Cliff Ice Tongue area at five hundred feet happened. We saw the surface formations and took some photographs of the area from the air. Then after a short sledge trip to take the kinks out of the snowmobiles to Cape Evans near McMurdo; the Nansen sleds, equipment, snowmobiles and supplies are loaded into the back of a ski equipped C-130 Hercules aircraft are flown to the deactivated Beardmore Camp on the Beardmore Glacier. After the flight landed at Beardmore Camp all the equipment and supplies were offloaded. We saw almost every day the contrails of the daily flight to South Pole by C-130 Hercules aircraft in the western sky. We had an airdrop of fuel during the latter part of the expedition made from C-130 Hercules aircraft. Then at the end of the season we, our equipment and our precious cargo of meteorites were picked up by a ski equipped C-130 Hercules aircraft at the deactivated Beardmore Camp. Then we were flown back to McMurdo from where we flew by ski equipped C-130 Hercules aircraft back to Christchurch from where we flew commercial airlines back to our respective homelands.

Their have been a sustained number of aerial photographic over flights and satellite remote sensing of many parts of the

Antarctic and there have been over flights of the Lewis Cliff since the late 1940's. The area of the Beardmore Glacier has been traversed since the early 20th century when Scott and Shackleton had active expeditions.

Under canvas:

The experience that I had in the Antarctica was one of living with five other expedition members in three small Scott Polar tents that were like pyramids. We each had our own snowmobile and sleds. We had cooking equipment that we used. If we scrunched up we could fit in a crowded manner all persons into one tent which we did for Christmas dinner. We would fill up the primus stoves outside seal them and then move them inside. In our tent we had 'Hoosh' or 'Hoosher' which is a concoction whereby you have one pot with food that is added at the appropriate meal time and leftovers are left in the bottom of the pot to freeze and then the next meal would add stuff to the gradually evolving meal. This is Hoosh and in the tent that I shared with the Japanese scientist we had the same pot with new stuff added for the entire season and even took it by sled to our last meal at Beardmore Camp. We slept in sleeping bags and used ziplock plastic bags to urinate inside of our tents during the night. Together in our tent I and Dr. Keizo Yanai ate 25 lobster tails over the field time. My tent-mate was Japanese and a noted Antarctic meteorite hunter who had worked with Dr. Cassidy since the early years of the American Antarctic Meteorite recovery Program.(Cassidy, 2003) We would get up in the morning and have hot chocolate and several packages of instant oatmeal and powdered milk to make up a good thick bowl of porridge. We would add a lot of sugar to the porridge. I would make a sandwich or a snack for lunch. We would spend the whole day out looking for meteorites and then come back to camp and then cook dinner usually in Hoosh. As we had little else to do we would stay in the tents and take our time cooking. When wind storms that gave to low wind chills came roaring in we would have to stay in our tent. At one very blustery night(the sun was up all the time)sometimes we would work outside for up to 18 hours a day if the weather was good; Dr. Cassidy came into our tent and told us that we should get into our sleeping bags with all of our clothes on because the wind was blowing faster then what the tents were rated at staying up at. Another day their was no wind and the temperature was so warm that we were able to go outside with just our clothes on without our jackets and toques and other accoutrements on. The Dutch scientist brought a portable windmill that we tested out in the field by setting it up and charging batteries from it. We would either drive to the Lewis Cliff Ice tongue, meteorite moraine or go on reconnaissance traverse such as to Goodwin Nunatak. At four o'clock in the morning the sun would be up in the southern sky and the sun would move around the sky in a circle.

We were very near the famous vertebrate fossil bed called Coal Sack Bluff that had much attention and field teams at it for several. It is ironic that they were so close within a distance of a few miles of over 1300 meteorites and in one field of recovered meteorite a two large JATO bottles were found which means that that site had been used for a jet assisted take of by presumably a C-130 Hercules aircraft.(Cassidy, 2003) This area for the plateau is a well traveled area of the continent and yet no meteorites were found until Dr. Cassidy(2003) came along.

Composition of field teams:

The way that the ANSMET, the American Meteorite Recovery Expedition, is to have at least half of the field members with at least one season in Antarctica. The remaining expedition members could be rookies as far as Antarctica is concerned. With each field team there would also be a crevasse expert which is someone that would be the guide or mountaineer who was also the only paid member of the expedition. We had a Dutch geochemist, a Japanese geologist, a field leader American geologist, a crevasse expert American, an Austrian geochemist and myself a Canadian geographer. The back of Dr. Cassidy's (2003) book Meteorites, Ice, and Antarctica lists all of the field members that went to the Antarctic with the ANSMET program over the time that he was the principle investigator of the American Antarctic Meteorite Recovery program and their institutional affiliations. The Antarctic meteorite recovery programs today are in the field and physically and mentally are the most challenging. Seasoned veteran Antarctic members are needed to guide the rookies.

Processes of concentration of Antarctic Meteorites:

Especially mapping of meteorites recovered has led to a font of information that has allowed a greater understanding of the processes of meteorite concentration on Antarctic blue ice stranding surfaces. Dr. Cassidy from an early point understood this and this has allowed a greater understanding of where new meteorite placers occur and can be targeted for recovery. In a simple analogy meteorites that lie in the upper layers of the ice cap and move to the edges and then move upward to be concentrated near the surface of barriers is similar in a way to how driftwood gets concentrated on the edge of a body of water near a barrier.

The following are the necessary three processes of ice concentration of meteorites. Detailed mapping has allowed for greater and more intricate theories of ice concentration.

Ice Moved:

Movement by ice is the primary way in which meteorites are theorized to concentrate on blue ice stranding ice placers. The three locations of meteorite stranding ice placers are as follows, "1/ Immediately adjacent to and the interior ice sheet (plateau) side of the nunataks or mountains such as at the Yamota Mountains. 2/ Immediately adjacent to and on the downhill (lee) side of front range nunataks or mountains such as at the Frontier Mountains. 3/ Surrounding major ice scarps, with or without exposed morainal material, such as Elephant moraine icefields."(Harvey, 2003, 100) These three proposed systems of categorizing meteorite stranding surfaces are- uphill, downhill and overflow.(Harvey, 2003,101) These three categories of location of recoverable meteorites have been honed over 30 years of dedicated meteorite recovery in Antarctica. The key constant ingredient of Antarctic meteorite stranding surfaces is that they are associated with a ice flow barrier. It is also interesting to note that the dates of fall to Earth, which can be measured can be substantially different such by up to several hundred thousand years. The meteorites that fell in different times might be recovered next to each other on a blue ice placer.

The first category of Antarctic meteorite stranding surfaces occurs up ice flow of a barrier- this is called uphill. The second category of Antarctic meteorite stranding surfaces occurs down ice flow of a barrier- this is called downhill. The third category of Antarctic meteorite stranding surfaces occurs on ice scarps barriers- this is called overflow. This is developing the earlier conveyor belt theory of Antarctic meteorite transport.

Remelting of blue ice areas:

During the Antarctic summer it does happen, that blue ice areas do melt a little bit on the surface. This can both cover up existing meteorites and expose some other meteorites. Sublimation can also occur on the blue ice areas. Two processes occur the first is exposure of new ice and new meteorites and the second is when there is melting of meteorite bearing ice which buries the meteorite and can transport meteorites underneath melted pools. It also can drain placers and new surfaces and meteorites can be exposed.

Melting of ice is one of two processes theorized as producing mass wasting of ice that reveals transported meteorites.

Cryoconite holes:

It has been proposed that meteorites sometimes sink down from the surface of the ice at a higher rate then the erosion of the blue ice. They form cryconite holes that move down as follows, "rocks are significantly denser than ice; they warm easily in the sun and under appropriate conditions will sink toward the bottom of the ice sheet."(Harvey, 2003, 111) Nagata (1982) says dense irons large meteorites sink more rapidly and disappear.

Aeolian moved:

There is evidence that meteorites move after getting to the surface of a stranding surface by the action of the wind. This is not clearly understood. This might have an effect on the average mass of meteorites that happen to be moved by the wind. In part of the stranding surfaces winds can get up to 200 km per hour. Meteorites that have been recovered in hot deserts have been moved around by heavy winds. The author believes that it is possible for meteorites to be moved over the East Antarctic Ice Cap entirely by wind transport or a combination of wind movement, then ice movement and then wind movement. Meteorites that are heavier would likely not be moved as much if at all as compared with meteorites with less mass.

Direct falls onto ice:

It is also possible that their might be a minority of the meteorites that are found that might have arrived at their present location by direct fall to the blue ice stranding surfaces. They would not be covered up by the build up of snow. There is basically most likely a combination at each site of the three methods of transportation which might confuse the meteorites that are recovered and mean that vastly different samples and fall time of meteorites happens. Dr. Gary Huss proposed that meteorite stranding surfaces have a larger portion of direct falls than has otherwise been proposed. This theory has problems when compared to the transport of meteorites theories. But there is a good chance that some of the recovered meteorites have been direct falls. It is not known as to what is the ratio of ice transported versus direct fall meteorites.

Katabatic Winds:

The process of exposure of blue ice meteorites means that winds have a distinct part to play in the exposure over time of meteorite standing surfaces. Ice moves in a conveyor belt process that encounters under ice or over the ice surface barriers that then there is an upwelling of meteorite bearing ice from the vast interior of the East Antarctic Ice Cap. "Furthermore, many Antarctic meteorite stranding surfaces are located in regions where katabatic winds converge, significantly enhancing their effect."(Harvey, 2003, 103)(Cresswell, 1988) These katabatic winds also play a factor in movement of meteorites and the process of weathering.

<u>Scouring</u>:

The physical force of the high katabatic winds can physically scour the snow and ice at stranding surfaces and eventually reveal over time meteorites that have moved coastward in the upper portion of the ice caps ice.

Melting:

The ice can melt and can form large pools. In one case Dr. Cassidy found an organism living in a melt water pool. Usually the melting is not that extensive in size but is a factor for erosion and appearance of submerged meteorites.

Sublimation:

This process is when the ice converts directly from solid to a vapor state and is a factor of how blue ice stranding surfaces could be eroded.

Transportation to and from placer sites:

There are two methods of transportation to Antarctic meteorite stranding surfaces. The first is by aircraft and the other is by surface traverse. Both these methods of transit to the Antarctic meteorite stranding surface can be used. It should be noted that one of the most dangerous and difficult portions of the field season is the insertion into the field camp by whatever means. On the last week of the expedition one of the skis on the front of my snowmobile broke because of over use and it was difficult for me to make turns or maneuver which made for a somewhat interesting traverse back to where we had been dropped off and then picked up again by C-130 Hercules at the end of the field season.

Using C-130 for transport:

The use of large ski equipped aircraft such as C-130 Hercules aircraft has distinct advantages. A team of six to eight members can be transported to anywhere in the continent of Antarctic with sleds, supplies and snowmobiles for each person. One disadvantage is that it needs a fairly secure place to land and take off from. This was the method of transportation for the six person team that I was a member of in the Antarctic at the Lewis Cliff Ice tongue. We were transported into the deactivated Beardmore camp on the plateau side of the Beardmore Glacier. Also C-130 Hercules aircraft can also para-drop resupplies of fuel and supplies anywhere in Antarctica. Again this was done during the season that I was in Antarctica in the austral summer of 1986-1987. We had fuel supplies para-droped to us in the last portion of our field season at our Lewis Cliff Ice Tongue camp. We had radio contact and use smoke canisters to mark the location where the pallets were set to be dropped over. Large ski equipped aircraft have the distinct logistical advantage of being able to put into the field anywhere in the Antarctic a full field team for an entire season with enough supplies and equipment for a full austral summer season at one locale.

Using Twin Otter for transport:

The main limitation of using Twin Otter aircraft or other small aircraft to place meteorite search teams into the field is that they do not have a large cargo capacity in the field. They have distinct contrary advantages. The first advantage is that they have a larger number of locations that they can fly into and position a team. They can take off and land in difficult terrain and can have shorter landing zones to fly into and out of. Because of this they are safer to use in the field. The second advantage is that they are more inexpensive and have greater logistical flexibility to be used. They are very advantageous to use for reconnaissance missions and allow several sites to be easily used. One example of this is when a British geological team discovered a meteorite in a previously untouched site and they flew their Twin Otter to the site where the ANSMET team was at and took back two member's of the ANSMET team with them for one week to search this new site where several meteorites were recovered and then the two Americans rejoined the main team when their week was up at the new stranding meteorite surface.(Cassidy, 2003) This shows the flexibility of using a small fixed wing aircraft such as the Twin Otter or camparible ski equipped small aircraft. The Americans use some Twin Otters at the current moment. Twin Otters can be flown into any part of the Trans-Antarctic mountain range where many American discovered and explored Antarctic blue ice stranding surfaces are along.

<u>Long Range surface traverses</u>:

This method of getting to Antarctic stranding surfaces is used by the Japanese Antarctic Meteorite Recovery Program. This means that they do not have a field team in the Antarctic searching for meteorites every year. Of all of the methods to place a team onto a blue ice meteorite placer this in the authors opinion is the most cumbersome but the Japanese do not have air-landing capability in their station in the Antarctic. They do not have the capability of a large heavy equipped aircraft such as the American C-130 Hercules aircraft. Ironically meteorites were discovered in the Antarctic in great numbers because the Japanese were 'on the ground.' This method of traverses could be used by other countries to traverse to potential blue ice meteorite stranding surface areas. This method could be emulated by other Antarctic nations that have stations around the Antarctic continent.

Robotic Contribution to Antarctic Meteorite Recovery Program:

It is only natural to consider the use of robots in the field in such a harsh and dangerous place such as Antarctica. The author's opinion is that robots have been and should be used in the field as adjuncts to human field work. NOMAD was successfully field tested and successfully identified meteorites in the field comparing them with the already present terrestrial material on the ice. I proposed in 1987(Mardon) the use of remotely piloted vehicles for meteorite location identification from the air. It is not a far step to from human controlled vehicles to the use and development of artificially intelligent air and surface remote sensing vehicles. Cassidy(2003) is to be commended on supporting the NOMAD initiative and encouraging them to field test the equipment in the Antarctic. Robotic vehicles are the backbone of the space programs robotic exploration of the solar system and will be the leading edge of the spear for the long term.

<u>Robot surface traverses</u>:

The Americans under Dr. 'Bill' Cassidy(2003) developed a robotic land vehicle called NOMAD. It was field tested on a blue ice stranding surface where meteorites that had been found were placed within a grid that was predetermined and then the NOMAD vehicle found each meteorite. I believe that with tele-video capability with transmission back to a state side facility rather then using artificial intelligence this would mean that samples that were meteorites but did not have the necessary spectral signature then they could be located and deduced. The author had a discussion when he was in the Antarctic in 1986(Mardon) where he suggested that every rock sample be collected from the blue ice zones by robots but it would have been to difficult to send back all of the terrestrial samples just to find the fraction of the surface materials that were meteorites. It would seem that samples in areas such as Meteorite Moraine at the Lewis Cliff Ice Tongue that are mixed up with large amounts of terrestrial material could be an appropriate area for robotic detection and recovery. In conjunction with manned expeditions it is a potential idea that should be examined in more depth. There is a real chance to improve robotic mission capabilities by field testing new technologies in Antarctic robotic meteorite recovery. It is possible that the Martian polar ice caps could have robots sent their to see whether meteorites could be recovered and then returned to Earth.

Robots on the ground may act as a supplement and potential aid to field teams in the Antarctic but a human presence is important to have human perspective and creativity. If humans had not been present when the Japanese expedition 'accidentally' found meteorites in the Antarctic an invaluable source of non-terrestrial samples would have not been developed in the Antarctic. A human presence in the field in Antarctica is a must for new things and ideas to be discovered.

Robotic aircraft for surveying:

When the author first proposed field use of portable remotely controlled aircraft the technology was either classified or not as developed as it currently is. Today aircraft have been miniaturized to the point where vehicles can be stored and used with equipment from a persons backpack. When I first proposed the use of Remotely Piloted vehicles it was very difficult to envisage in the field as the launching equipment and control equipment was very cumbersome. Today technology has advanced to the point where one person on a snowmobile or on foot could carry all of the equipment for the use of field aerial photography. It would also be of use to examine the potential traverse routes of reconnaissance teams that are going through areas for meteorites for the first time. Images of the field could be beamed back directly to a station in the United States of America where they could be looked at by a team. Traverse routes through crevasse fields could be selected as it would be easier then try to find the safest route through by selecting routes from the top of a snowmobile. Potential meteorites as I have previously proposed in 1985 (Mardon) could be detected.

Analogue Terrestrial comparisons with Antarctic Meteorite Recovery locales:

Antarctica is the world's harshest and driest desert. Like in other deserts meteorites are found in the Antarctica. Unlike hot deserts the cold desert of the Antarctic substantially concentrates meteorites in stranding surfaces due to Aeolian and ice conveyor belt mechanisms. It is possible that the Martian polar ice caps might have concentrated and preserved meteorites in a similar way as East Antarctic Ice Cap has concentrated and preserved meteorites.(Cassidy, 2003) Other locales outside of Antarctica such as Pleistocene northern latitude ice caps might have concentrated meteorites and preserved them. The high Arctic might also have had a process of concentration of meteorites. Greenland might still be as active in terms of concentrating blue ice meteorites. Micrometeorites might also be concentrated and potentially recovered.

United States expedition to the Greenland Ice cap to recover Meteorites:

Dr. Ralph Harvey led a 2004 expedition to Greenland to see whether it would be possible to recover meteorites from the Greenland Ice Cap that would have similar methods of concentration of meteorites to what currently happens on the East Antarctic Ice Cap. Greenland is the only other very large ice cap in the world. The theory is that the process of concentration by ice movement, Aeolian movement and direct falls to blue ice areas occurs in the Greenland Ice Cap. Cosmic dust and micrometeorites might also be concentrated in dust bands that could be recovered from blue ice that has dust bands in the exposed blue ice zones. Potential falls recovered from Greenland might indicate different groups as the process of ice concentration is different in Greenland. Meteorites have been recovered from northern Greenland. Several large meteorites were recovered during the early 20th century during several Arctic expeditions to the Arctic.

If meteorites are found in Greenland blue ice placers then it would be possible to maybe consider that meteorites might also be found in other stable ice caps or left as post glacial debris in northern latitudes.

Pleistocene deposits of meteorites in Canada:

Substantial deposits of meteorites have been found in hot deserts due to direct infall to the locations where they were found later. In Antarctica there is a portion of meteorites found on blue ice stranding surfaces that have been theorized are from direct infall to where they are later found. Much of northern Canada was glaciated during the last major period of glaciation. If meteorite concentration by ice movement occurs in Antarctica due to ice movement and also post exposure Aeolian movement then the author has proposed that moraines in Northern Canada and Alaska could contain concentrated placers of meteorites, micro-meteorites and cosmic dust. It is possible that their could be meteorite concentrations in previously glaciated portions of Europe, Asia and North America of a similar concentration into dense placers as is found in the Antarctica around the East Antarctic Ice Cap.

In the Arctic there would have likely been stable ice movement as what happened in Antarctica. The high Arctic of Canada could have meteorite placers where blue ice would have been created as what happened in the Antarctic. Meteorite concentrations that have been recovered occur near or at barriers where there is blue ice and katabatic winds. The Rocky Mountains of Western Canada might fit the bill of being a potential barrier for ice movement and had katabatic wind. Although the author is not familiar with current and previous glaciation in the high Canadian Arctic it would seem that meteorites could be concentrated, preserved and also might be recovered in the High Canadian Arctic.

With global warming the ice caps are receding in the Canadian Arctic it is likely that meteorites could only be found in the terminal moraines of ice caps and glaciers. Direct falls of meteorites could be found on the tundra and formerly glaciated islands and mainland in the High Arctic.

Columbia ice field in Canada:

The Columbia Ice Field has is a remnant of the last major glaciation of the Canadian Rocky Mountains. The question that the author has wondered about since he went to the Antarctic is whether the Ice Field could concentrate meteorites like what occurs in the Antarctic on the edge and in the Trans Antarctic Mountain Range. When the idea was published in the 1980's there was a little bit of interest but their was no real reaction to idea. Currently the ice Field is not stable and has been decreasing in size. The author believes that part of the success of the Antarctic meteorite recovery program is that there is little clutter of terrestrial rocks as compared to other terrestrial locations. It is possible that in the moraines of this ice field- the Columbia Ice Field their might be concentrated meteorites, micro-meteorites and cosmic dust. It just might mean for one season or for Parks Canada geologists to be aware of the potential find of non-terrestrial rocks.

It took a long time in the Antarctic before scholars realized that blue ice areas were a bonanza for meteorite recovery. It is possible that this ice field might show the same pattern of meteorite concentration as the East Antarctic Ice Sheet.

Recovery of meteorites in Canadian Arctic:

Like other deserts the high Canadian Arctic could be a source of meteorites. One difference between Antarctica and the

Arctic is that in the Arctic it is occupied by the Inuit. The people that live in the Arctic could assist in recovering meteorites if the Inuit had a short training as to what a meteorite in the field would look like. The last Pleistocene glaciation in the Arctic might have concentrated meteorites. Also their would be the chance of recovering direct infall locations and their might be some Aeolian concentration. As has been previously suggested it might mean recovery of samples if short workshops occurred in Inuit communities on what to look for in a potential meteorite that was come upon in the field in the Canadian Arctic.

Pleistocene deposits of meteorites in Russia:

Major portions of the northern Russia and Europe were glaciated and it could be that meteorites were concentrated in the same method as they were concentrated in the Antarctic by direct fall, ice movement and aeolian movement. It is possible that they could exist in the ice caps in the northern Russian arctic islands. They would be preserved as they are in the Antarctic. The author does not know of any Russian attempts to search for meteorites deposited during the Pleistocene in northern Russia. Meteorites have been found in the Canadian Arctic and in northern Greenland. If the Russian government gave a bounty for the recovery of meteorites then the local inhabitants of northern Siberia and northern Russia up to and including the Arctic archipelago then meteorites could be looked for as other minerals are recovered in the vast Russian taiga. This method of using the locals to bring in potential meteorites was pioneered during the depression by the Berringer the American geologist.

Meteorites lack terrestrial contamination:

The volume of frozen generally uncontaminated by biological activity and only weathered by the ice and the cold means that samples recovered are some of the least uncontaminated samples that have been recovered. Because of this it also means that the samples recovered are likely a more representative sample of meteor flux rates impacting the Earth over the last two million years. Different Antarctic stranding blue ice surfaces have different periods of time that the meteorites found there that they represent. One incident was when we were at Meteorite Morraine one of the team was relieving themselves and they found a meteorite that way. Hopefully that meteorite did not get terrestrial biological contamination!

There can be substantial weathering of the meteorite once it appears on the surface of the ice. It has been theorized that the snow crystals act as sandblasters in the mid-winter wind speeds of up to 200 kilometers per hour.

The Americans ship the meteorite samples back to the United States in a frozen state and when they open up the bags in Houston at the Johnson Space Centre they do this in a nitrogen atmosphere to reduce contamination when the samples are thawed out. The samples are examined and stored in the same manner as the geological samples brought back from the Moon by the Apollo missions. The author does not know how other nations reduce the chance of contamination in the meteorite collection and archival procedures. This reduction of contamination is important in the study of meteorites.

How meteorites are archived and used after recovery:

The system for archival that is the most comprehensive after the meteorites have been sent back from the Antarctic is the American system. They use clean bags and ship back the samples frozen and then examined each sample at the Johnson Space Centre and then once the use of the sample has occurred the samples are sent to the Smithsonian where they are archived for posterity. (Cassidy, 2003) The meteorites are opened in a nitrogen atmosphere and thawed to room temperature. The Americans have a committee called the Meteor Working Group that decides which researcher and which research proposal samples from meteorites are allocated to. The Americans do a mineral classify of each separate sample that is sent back from Antarctica. Each nation has different protocols and procedures for the preservation and scientific use of meteorites.

After the American Meteor Working Group decides that a specific meteorite is redundant and no new substantial research will occur the meteorite is sent to the Smithsonian Institute for archiving and preservation. The Americans have provided any domestic or international researcher with a meteorite sample after scholarly peer review by the Meteor Working Group of the proposed use of the meteorite sample. They have not played politics with the samples that were brought back from the Antarctic. Certain samples of meteorites such as the unique ones, lunar meteorites and Martian samples should be viewed as the heritage of humanity and all qualified and peer reviewed research should be able to get access to samples for research.

Helicopters being used to find meteorites:

When Dr. Cassidy first received funding he realized that it was important to start out in areas that were close to the logistical hub of McMurdo Station. In his book he says that, "I found a number of such areas(blue ice) within helicopter distance of McMurdo, and these seemed good places to begin our search."(Cassidy, 2003, 24) The first blue ice areas to be examined by the Americans were blue ice areas that were near the logistical capabilities of helicopters from McMurdo. This range of rotary wing aircraft was a key factor early on before the use of C-130 Hercules aircraft came into their own.

<u>Side mounted video cameras</u>:

The author proposed in 1986 that helicopters be fixed with either 70mm or video cameras and record the areas that they fly over and then back at camp or at McMurdo Station have someone view the camera to see whether they could distinguish meteorites. As far as the author knows this has not ever been attempted. One downside is that most of the blue ice fields near the American McMurdo station have been systematically searched and now it is necessary to go further a field to sites that can only be accessed by fixed wing aircraft instead of rotary wing aircraft.

Persons flying:

Early on in the ANSMET program the helicopter would fly low over a blue ice area and then a person or two would look out of the side and front of the helicopter and stop and land when a potential meteorite was sited and then run over to the meteorite. This was told to the author in the field in Antarctic by the expert mountaineer who did this and actually found several meteorites using this method. It would obviously be a lot safer to use remotely piloted vehicles then to use helicopters.

Aerial Photography Contribution:

From the beginning of the American meteorite recovery program aerial photography and remote sensing have been important in the selection of potential reconnaissance sites. "The United States Geological Survey has a tremendous library aerial and satellite photos of Antarctica, and I became a frequent visitor to their archives located in Reston, Virginia. The satellite photos were on filmstrips."(Cassidy, 2003, 22) Dr. Cassidy from early on understood the reconnaissance value of using aerial photography to select potential blue ice meteorite placers. Tony Meunier who worked for the United States Geological Survey went down to the

Antarctica before I went to the Antarctic to set up benchmarks for surveying the Lewis Cliff Ice Tongue meteorite deposits. Dr. John Hobart Roscoe wrote a seminal dissertation on the field of Antarctic aerial photography in the 1940's that was based on his research in the Antarctic. The author has had correspondence with Dr. Roscoe who is currently retired in California since the mid-1980's and supported my membership on the 1986-1987 ANSMET expedition to the Antarctic. He gave the compliment that the idea that the author proposed was an idea that he himself would have come up with if he had heard of the Antarctic meteorite endeavor.

Two large meteorites might show up on regular aerial photography:

Theoretically it might be possible to view large meteorites on archived photography. Aerial photography of blue ice placers goes back to just after World War II. The size of the meteorites would have to be about a square meter for the meteorite to show up on archived aerial photography. When the author visited the United States Geological Survey in Reston, Virginia he examined archived photographs of the Lewis Cliff Ice Tongue going back to the 1940's and the author believes he saw one meteorite that was very large on a 1967 archived aerial photograph. There was a dot on the film in the north west portion of the Lewis Cliff Ice tongue where a large meteorite was recovered by ANSMET. The author has contacted the current leader Dr. Ralph Harvey of the ANSMET program to see whether it would be possible to get funding for research into the recovery of archived film showing large known mapped meteorites. Because the author is not a United States citizen it is difficult to get funding to do research on American aerial photographic films. The author has not investigated other nation's aerial photography of the Lewis Cliff Ice Tongue. As I previously stated the Lewis Cliff Ice Tongue is part of the Antarctic that has been visited on the ground and flown over by air frequently over the years since the end of the Second World War. It was very near where Scott ascended the Beardmore Glacier to his tragic expedition to the South Pole.

Archived aerial photography:

When Dr. Cassidy (2003) originally began to apply what was found in the Japanese discovery of meteorites then he applied the concept of using aerial photography to look for blue ice areas near to McMurdo Station that could be accessed by helicopter logistically. This distance is 185 miles. From the very first the Americans used the technology of archived aerial photography to find potential blue ice meteorite standing surfaces. Very large meteorites might be able to be detected using archived photography transects from before the meteorites were recovered. We already know where the meteorite lay on the ice and it would be very simple to cross reference those known coordinates with the same coordinates of aerial over flights. Another interesting research topic using aerial photography is to do detailed measurements of the snow cover and its distribution over a small blue ice area such as the Lewis Cliff Ice Tongue to see whether there is a fluctuation of the snow cover. This might lead to the possibility of finding out whether the meteorite stranding surfaces might in the near future have new portions of the ice uncover from snow. Their has been some indication that subsequent surveys of meteorite stranding surface have meant tht new meteorites have been found in areas that have been intensely searched. Snow cover over time on these meteorite stranding surfaces might help searching those placers again.

Current aerial photography:

"Blue ice areas are commonly visible in satellite imagery and aerial photography, and reconnaissance for new meteorite recovery sites usually starts with a survey of available imagery." (Harvey, 2003; Lucchitta, 1987) "In aerial photos one can often see details that indicate ice-flow directions."(Cassidy, 2003, 23)

Some examples of how aerial photography is depicted in scholarly communications are as follows, Harvey(2003, 119) says Lewis Cliff Ice Tongue shows a mosaic of vertical aerial photographs with an overlay of recovered meteorite locations. Another referenced article is as follows, Harvey(2003, 124) shows Foggy Bottom/ Goodwin Nunataks region of mosaic of aerial photography with meteorite locations overlaid. In Dr. Cassidy's seminal description of the Antarctic meteorite effort it is said that also Cassidy(2003, 300) photograph shows morainal ridges north of the Lewis Cliff Ice Tongue. In the above photographic representations it should be noted that they are all seminal to the scholarly threads of the article and book that were examined. They spatially represent information that otherwise would be difficult to depict using the written word.

Remote Sensing Contribution:

Another geographic tool that is being used is satellite imaging. Dr. Cassidy said, "I was able to search large areas of the ice sheet by remote sensing. I looked not for meteorites, because they would be to small, but for large patches of ice because one wouldn't expect to find meteorites in deep snow."(Cassidy, 2003, 22) The entire reason of going to the Antarctic was to find meteorites for the ANSMET team and it was obvious to Dr. Cassidy that blue ice zones were a potential bonanza for finding meteorites. That was an understatement. From an early point in his Antarctic career Dr. Cassidy later said of his eureka moment that, "I learned that many very large exposures of ice exist in association with the Trans-Antarctic Mountain Range."(Cassidy, 2003, 23) This was done off of earlier LANDSAT images of the blue ice areas on the Trans-Antarctic Mountain Range.

Satellite imaging contribution:

"Blue ice areas are commonly visible in satellite imagery and aerial photography, and reconnaissance for new meteorite recovery sites usually starts with a survey of available imagery." (Harvey, 2003; Lucchitta, 1987) Harvey(2003, 101) use maps with overlay from LANDSAT 7 of Meteorite Hills icefield. Cassidy has locations of Allan Hills and Elephant Moraine wite from satellite photography.(Cassidy, 2003, i) "But existing satellite photos gave us the ability to identify ice patches in all parts of the continent, and there were more distant places that we aspired to visit." Cassidy, 2003, 57)

Aerial remote sensing contribution:

Their have been over-flights of blue ice areas but none dedicated to meteorite recovery. These over-flights contribute to our understanding of the blue ice meteorite stranding surfaces. The authors opinion is that technology of satellites has advanced to such an extent that remote sensing using remote sensing aircraft can be covered by satellites. Although it should be noted that pixel sizes of civilian satellite remote sensing is not yet as good as that that can be acquired from aircraft mounted remote sensing systems. Such technologies that could be mounted on aircraft to sense meteorite stranding surface are SLAR Side Looking air radar. There is used of ice penetrating wavelengths to map out the under ice rock topographic radar. It is possible to use previously taken remote sensed images to view blue ice. Radar distinctly shows up on radar images when the ice is lying on the surface.

Potential use of radar on blue ice areas:

Radar remote sensing has been used for along time in the sensing of bodies or water and ice and has been used to sense potential blue ice areas as the following illustration from Harvey demonstrates. Harvey(2003, 123) uses RADARSAT of Walcott

Neve and Foggy Bottom area showing overlays of ice movement. Radar shows blue ice areas as dark areas and was used in identifying potential areas for reconnaissance. This is an excellent tool for the selection of potential blue ice stranding surfaces. Remote sensing is used in concurrence with aerial photography, mapping and talking with field scientists who might have found one or two meteorites on some exposed blue ice potential placers.

Cartographic Contribution:

Maps and mapping have been a major tool in developing major tools in assisting the understanding of the concentration of meteorites in Antarctica. "Regional maps, including the Yamato, Belgico, and Sor Rondane Mountains in Queen Maud land" is where the Japanese teams have focused and mapped.(Cassidy, 2003,289) The Japanese do not map as comprehensively or in as much detail as the American program. The author does not know how the other Antarctic programs map out meteorites in the Antarctic. This map was example of how Antarctic sites are located on maps for use in almost every paper that examines Antarctic meteorite recovery. From the very beginning even from before posting of maps on the internet maps of location of meteorites from specific sites such as the Lewis Cliff Ice Tongue was conducted. The Americans have something that used to be in hard copy and is now on the internet called the Antarctic Meteorite Location and Mapping Project(AMLAMP). The website to this site is www.lpi.usra.edu/research/amlamp/intro/intro.html It is a comprehensive of the mapping that is directly done by the ANSMET. It is a very, very comprehensive cartographic database. Since the beginning the Americans have realized the need for keeping and collecting information that is related to field occurrence of meteorites rather then just focusing on the geochemistry of meteorites. If this information specifically based on the cartography of meteorite occurrence it would not be possible to have a greater understanding of meteorite stranding surfaces. This has also led to better and more successful meteorite surface traverses. This is the same commitment to detail where the Americans catalogue each sample even if they are similar to many previously samples recovered this commitment to detail has given us and understanding that would not necessarily be their if the Americans had followed the less detailed analysis of the Japanese Antarctic Meteorite Recovery Program.

How sites are selected:

General use of maps in conjunction with aerial photography is how new sites are selected. Harvey (2003, 118) uses maps to show ice processes and meteorites. Cassidy(2003,10) shows locations of major meteorite concentrations in Antarctica. Initially blue ice areas were selected near to the McMurdo Station so that the field teams could get to the blue ice meteorite stranding surfaces by helicopter from McMurdo. Soon after the start of the ANSMET program sites were selected that were further afield outside of the McMurdo Station helicopter support range. Then sites would be examined where previous meteorites were discovered by other non-meteorite recovery programs or where the blue ice was exposed and a reconnaissance party was sent to examine the blue ice surface for meteorites.

Surveying and mapping meteorite locations:

In my personal experience mapping of meteorites does take time but it does give a window and data into how meteorites are concentrated. Cassidy (2003,297) shows different concentrations of meteorites near the Lewis Cliff ice Tongue due to where there ice catchments basin is. Each meteorite that is found on a systematic search is marked and then surveyed from benchmarks that are predermined and located accurately by Global Positioning System(GPS).

United States of America mapping of meteorite placers:

An example of the mapping is that in Cassidy's book (2003, 294) where he has mapped out the Lewis Cliff Ice tongue and the features surrounding it. The American's since the early phase of the Antarctic meteorite recovery program have focused on keeping cartographic records. The Americans have something that used to be in hard copy and is now on the internet called the Antarctic Meteorite Location and Mapping Project(AMLAMP).

The website to this site is www.lpi.usra.edu/research/amlamp/intro/intro.html

Contribution of the writer:

The author will describe the method and process of exploration that is used to find Antarctic meteorites using geographic techniques. Geographic techniques have been used from the beginning when Japanese Antarctic explorers first recovered meteorites. The author will examine how geography has been used and will be used. An example is the mapping out of meteorite recovery locations within each of the placers such as the Lewis Cliff ice tongue complex. The author flew in an over flight of the Lewis Cliff ice tongue, Antarctica and examined the last 50 years of aerial photography and remote sensing of the Lewis Cliff ice tongue, Antarctica in Washington, D.C.(Mardon, 1989) The author was the first geographer to go on an Antarctic meteorite recovery expedition in 1986. (Mardon, 1988b)

The following two proposals for new Antarctic meteorite recovery programs were intending to increase the number of countries that look for Antarctic meteorites on the meteorite placers that they are found at.

<u>Proposed Russian Antarctic Meteorite recovery program</u>:

The author was the first to propose that the Russian scientific Antarctic Program set up an Antarctic meteorite recovery program.(Mardon, 1992a) The Russian program could use surface traverses to recover meteorites. The United States of America Antarctic meteorite recovery program uses fixed wing aircraft to get put recovery teams into prospective placers. A proposed Russian

Antarctic meteorite recovery program could save money by traversing on the surface from a major station such as Mirny to blue ice areas near the Trans-Antarctic Mountain range. (Mardon, 1998a) This would be similar to how the Japanese Antarctic meteorite recovery program recovers meteorites. Unlike the American program, the Japanese have extensive surface traverses to meteorite placer zones and then search blue ice areas for meteorites. As far as the author knows there has been no interest in Russia relating to setting up an Antarctic meteorite recovery program. The proposed Russian Antarctic meteorite recovery program could cooperate with the Japanese Antarctic meteorite recovery program by using the existing Russian bases in the Antarctic.(Mardon, 1998a) Also it would use the Russian knowledge of long distance surface traverses in the Antarctic. Vostok station could be used as a potential staging post for a Russian meteorite recovery program. During the traverse to Vostok station it might be possible to search at least at a reconnaissance level for potential meteorites at blue ice placers.

The following section is similar to the previous section on having a Russian Antarctic meteorite recovery program except is a proposal for a new Antarctic meteorite recovery program for a nation that does not have an Antarctic meteorite recovery program.

Proposed French Antarctic Meteorite recovery program:

The author was the first to propose that the French Antarctic Program set up an Antarctic meteorite recovery program.(Mardon, 2003) What was proposed was that the French

Antarctic program set up an Antarctic meteorite recovery program. Like the proposed Russian Antarctic meteorite recovery program that was proposed in 1992 by the author (Mardon, 1992a) the proposed French Antarctic meteorite recovery program would travel to potential blue ice meteorite placers by surface traverses. This would be the same method of traverse as the Japanese use to get to blue ice meteorite placers. Their has been no feedback from the French government program since the author published his idea at the Meteoritical Society meeting in Germany in July, 2003.(Mardon, 2003) Their has been French scientists that have gone with the American ANSMET program to the field in the Antarctic.

The following section relates to the proposal by the author in 1985, in writing to the principle investigator of the United States Antarctic meteorite recovery program, Dr. Bill Cassidy to use aerial photography to directly detect surficial Antarctic blue ice meteorites from the air.

Use of aerial photography to directly detect meteorites:

In the spring of 1985 the author wrote a letter to a researcher at the Smithsonian Institute which eventually resulted in membership as a team member of an Antarctic field expedition 170 kilometers from the South Pole at the southern edge of the Beardmore glacier around the Lewis Cliff ice tongue. (Mardon, 1985) One C-130 Hercules aerial over-flight was conducted where photographs were taken. Infrared and regular photographs were taken.(Mardon, 1987f) Later in the field ground truth photographs were taken of selected meteorites and other rocks from the surface of the ice. Sadly it was concluded after the end of the season that it was concluded that it was not possible to distinguish between terrestrial rocks and extra-terrestrial rocks on the surface of the blue ice moraines. (Mardon, 1987c) The meteorites in many cases have black fusion crusts over the surface. This is from the heat of the entry of the meteorite through the atmosphere. Meteorites are found on the surface of the blue ice. This is partially because it would be very difficult to find meteorites if they are below the surface of the blue ice. The other idea was to use helicopters with cameras on the side of the air vehicle to fly over potential sites to be examined back in the United States or at the McMurdo Sound station.(Mardon, 1988a)

What is able to be deduced is that the blue ice areas can be assumed using criteria that is deduced from where meteorites have been previously found.

The following was an idea that was first voiced in 1987 in Houston, Texas at the Lunar and Planetary Institute Science Conference. The next year a metal detector was taken to the Antarctic.

Use of portable hand held metal detectors at blue ice Antarctic meteorite placers:

The author proposed in 1987 that portable hand held metal detectors be taken to the Antarctic blue ice meteorite placers. In the 1987-1988 season hand held metal detectors were tested but were found that their use did not substantially improve the detection of meteorites compared to conventional eyesight detection methods. They might theoretically have a use in heavily cluttered moraines such as 'meteorite moraine which was a very dense field of meteorites with very many terrestrial rocks. It would be very useful in detecting metallic meteorites and distinguishing between terrestrial and some extra-terrestrial rocks. It might be useful for the use of field teams going over very cluttered moraines that have previously been worked on.

The following is an idea that first came out over a year after the author returned from the Antarctic. It was to use archived aerial photography to detect already recovered meteorites on archived aerial film.

Detection of meteorites from archived aerial photography:

The idea was first presented by the author at a Association of American Geographers meeting that was held in Baltimore, Maryland, United States of America(Mardon, 1989) several years after the author was a member of the Antarctic meteorite recovery expedition. Their have been almost continual over-flights of the Lewis Cliff Ice Tongue since the 1940's the area also was traversed by Scott's party in the tragic expedition to the South Pole in the early twentieth century. There was one very large meteorite, weighing over 30 kilograms that was found on the Lewis Cliff ice tongue before the author was a member in 1986- 1987 of the meteorite recovery expedition. Theoretically this meteorite should be recoverable as an image off of some of the earlier mapping aerial over-flights taken by the United States of America's Antarctic program. (Mardon, 2004b) The size was over one square meter of surface of the meteorite. Some other meteorites were smaller but with enhancement would be able to be detected. Historical civilian satellite images would not have the pixel resolution to work as having the ability to detect already recovered meteorite. (Mardon, 2004b) Many of the meteorites that were found the year that the author was a member at the Lewis Cliff Ice tongue were several kilograms and were about the size of a mans fist.

The following is a description of potential analogies to where ancient meteorites could be found in previously glaciated northern latitudes. This is especially in North America.

<u>The Deposition of Pleistocene Placer Deposits Located in North America</u>:

By understanding the areas that meteorites are found in the Antarctic and deciding upon criteria to correctly predict meteorite recovery placer it is also possible to predict ancient placer sites in North America and in Russia.(Mardon, 1988c; Mardon, 1988d) Theoretically geologists should be able to recover meteorites in northern latitudes that were previously glaciated that would have similar depositional processes to those that are found in the Antarctic where meteorites are concentrated in blue ice upwelling areas near mountain ranges. The Antarctic meteorite recovery program is having an expedition to the Greenland Ice cap is this northern latitude summer as a reconnaissance expedition to try to find meteorites. Meteorites might theoretically be found in previously glaciated areas as they might be concentrated due to ice movements. Many meteorites are found in arid deserts because the meteorites are easier to spot and are not subsumed into surrounding terrestrial rocks, soil and plant life. This makes it more difficult to see meteorites.

The following section includes the use of robotic vehicles to recover meteorites. Field testing has not occurred for robotic air vehicles but has occurred for a surface traverse vehicle.

Use of aerial robotic vehicles to detect Antarctic surficial blue ice meteorites:

The author was the first to propose the use of robotic air vehicles to directly detect surficial meteorites using robotic vehicles in 1988 at an Association of American Geographers meeting in Baltimore, Maryland, United States of America. (Mardon, 1988e) In later years a robotic surface vehicle was tested doing a surface traverse that was in cooperation with the French Antarctic program. It was a successful experiment. The machine was able in the field to distinguish between terrestrial and extra- terrestrial samples. The proposed drone air vehicle was well received but due to budgetary limits did not go further then the proposal idea.(Mardon, 1988e)

The next section examines the idea to use civilian and military radars to detect fall locations of meteorites.

Proposed use of radar systems for the recovery of meteorites:

The author proposed that radar systems be used to find the exact fall locations of meteorites in Antarctica and in Russia and the United States of America.(Mardon, 1997) This concept is practical and only needs funding to convert data that is already collected in the northern hemisphere but in Antarctica would need radar stations to be set up. It would be easy to find meteorites when they land because there would be little terrestrial rock clutter to interfere and confuse any meteorite recovery effort.

The following concluding section covers the conclusion of this essay which is that geographic techniques have made contributions to the international effort to recover meteorites in the Antarctic since the 1970's.

Future Directions and Implications:

The stunning number and quality of Antarctic meteorites have given the opportunity to give geochemists that analyze samples from all around the inner solar system and from even beyond a chance to gain an incredible insight into the geological history of our solar system. Over 25,000 meteorites have been recovered from the Antarctic since Douglas Mawson's Australian expedition recovered the first Antarctic meteorite early in the 20th century in Antarctica, specifically in 1912.(Bayly and Stillwell, 1923) The 25,000 meteorite samples are a precious global resource and the

Americans especially have allowed researchers from all countries of the world to have access to them for researchers from all around the world.

Mining the Asteroid Belt:

The author has proposed that an eventual non-terrestrial source of strategic mineral resources could come from the asteroid belt.(Mardon, 1990) The only material geological samples from the asteroid belt are from meteorites. They can compare the spectral signature of the meteorite samples and then compare it to the spectral signature of the large asteroid in the asteroid belt. The asteroid belt is closer to low Earth orbit then the surface of the Earth this being based on the Delta Velocity force that is required to get to low Earth Orbit from the Earths surface compared to the Delta Velocity force needed to get to Low Earth orbit from the Asteroid belt. Long term resource source and distribution and development of inner solar system geological resources depend on an understanding of the chemical and geochemical nature of objects that would be mined in the inner solar system especially the asteroid belt. Terrestrial sources of strategic minerals is decreasing and ultimately the only new source of new mineral deposits for Earth and Earth orbit is the asteroid belt. The gravitational well from the Moon's surface to low Earth Orbit is also more costly then from the Asteroid Belt.

The Trojan asteroids are also a potential source of materials. Also we would like to not have all of eggs in one basket in case of a cosmic disaster on the Earth. Within the next several centuries space could be utilized for the development of resources that could be used to develop build energy producing systems such as Solar Power Satellites that could beam energy back down to the Earth from Orbit. The infrastructure in Orbit to develop Solar Power Satellites would need substantial construction materials that might be acquired from the Asteroid belt. It might seem very speculative but meteorites are a 'poor man's' space probe and with the 25,000 distinct samples that have been recovered it would seem that we have just to today enough separate samples to do geochemical analysis for several generations. With the advent of Antarctic meteorites the mass of material is the problem the samples are being recovered quicker then they can be looked at by scientists in detail. Space to any extent will only be colonized when there is a need for the resources of the inner solar system for mans push out to find new planetary homes.

Will become eclipsed by non-terrestrial sample return missions:

Proposed sample return missions from Mars, the asteroid belt and returning to the Moon. The collection of samples fallen meteorite samples will be at the forefront of research contributions to our understanding of the geology of the inner solar system and its varied geochemical vistas. With the emphasis of the American and other nations space programs on manned missions at the expense of robotic missions to more scientifically interesting scientifically locales in the solar system. If the same rate of recovery of meteorites from the Antarctic continues then it is possible that by the time substantial amounts of samples are returned to the Earth our understanding of the geology of other small and large bodies in the solar system will be substantially developed. Greater impetus has been given to the sample return missions to Mars and greater interest in Mars surface geochemistry because of the fact that evidence of life was potentially indicated within an Antarctic meteorite. The importance of Antarctic meteorites will not be supplanted until geologists are living and conducting science on the various large and small bodies of the inner solar system. Meteorite samples that are sold on the private market have great current value. It is comparable to the extreme value during the Early Enlightenment in the Netherlands for large sea shells that had been brought back from the other side of the planet. Over time they went from being very valuable to being a tourist trinkets that could be bought on any street corner for a few dollars. This is the same situation as there is for Antarctic meteorites and the potential for non-terrestrial sample return missions that would bring back samples to the Earth. Today they have inestimatable value but with time and more space missions will become less valuable scientifically and monetarily.

Meteorite recovery teams as analogy of international long term manned space missions:

The small teams that were started by the Americans have been since the very beginning international in nature which in some ways is unique for the American Antarctic program. They have at least half of the members of the team who are experienced Antarcticans. They pair up an experienced and an inexperienced member of the expedition. They also were all scholars at various levels of their careers including sometimes field scientists and sometimes laboratory scientists. Also they would have at least one 'dogo' would be the field guide. All would have to speak English at least with marginal fluency. For the future will this be the type of scientific expedition composition that will be found for scientific missions to the Moon, returning hopefully again, and beyond. Antarctic expeditions are and can be a model for short and long term manned space missions. The physical and mental challenges at least mimic what it would be like in space for manned missions.

Canadian Antarctic Program:

Canada has a robust Arctic scientific research program that encompasses research in her Arctic territories. Canada is the only polar nation, whether that be southern or northern polar nation that does not have an active Antarctic science program in the Antarctic. There is some reticence in the Canadian scholarly community that their should not be an Canadian Antarctic Scientific program because there is a feeling in that community that it would take scarce funding away from existing research programs in the Arctic. Canadian logistics experts and Canadian companies already serve other nations Antarctic programs under other nation's flags. Icebreakers that are not used in the Northern hemisphere during the Canadian winter could be redeployed to the Southern Antarctic waters in support of Canadian and other nations scientific programs. The Antarctic is the thermometer for the world and effects of climate such as global climate change can be predicted and measured in the Antarctic readily. It should be noted that the changes in the southern hemispheres high atmosphere ozone layer. Canada is in a position to use expertise and data from her northern climes and comparable data sets and research endeavors.

In summary Canada should have an Antarctic program within which their should be a Canadian Antarctic meteorite recovery program.

Specifically proposed Antarctic meteorite recovery expeditions:

Canadian aircraft company Ken Borak Air based out of Calgary already supports the United States of America Antarctic program by flying in Twin Otter Aircraft to the Antarctic. This company flew a mercy mission to the South Pole to evacuate a sick member of the over wintering crew at the United States South Pole Station. Canada already has the domestic capability to mount a three person field team to the portion of the Trans-Antarctic Mountain nearest the Antarctic Peninsula. Two aircraft could fly in the supplies and equipment necessary for a reconnaissance of potential meteorite placers already noted or examine new sites that have not been visited. It would be easy to find one experienced field guide and two field scientists to cover potential placers by snowmobiles that were flown in by aircraft. Since Canada does not have any station this is the reason for a two plane flight in and then after a month a two plane extraction. I would estimate that a mission would cost 150,000 for a team of three to go in for one austral summer season. The Patrick Hills might be a good location to first go to. It would of course be important to cooperate and coordinate with other Antarctic scientific countries. In small teams the Americans already use Twin Otters to fly into areas were a limited number of people travel to reconnaissance for meteorites.

Importance of meteorites to science:

It might seem very speculative but meteorites are a 'poor man's' space probe and with the 25,000 distinct samples that have been recovered it would seem that we have just to today enough separate samples to do geochemical analysis for several generations. With the advent of Antarctic meteorites the mass of material is the problem the samples are being recovered quicker then they can be analyzed in detail.

In a recent paper in 2003 Dr. Ralph Harvey the current principle investigator of the ANSMET program stated that, "The Earth-Moon system, a significant gravity well, careens along through these streams and clouds of Asteroidal debris, gathering some small portion of it onto the Earth's very own glassy windshield, the East Antarctic Ice sheet. But the journey of the meteorites is not over yet; gravity is not yet done with them. As the ice sheet sags and flows downhill under its own weight, it carries its cargo of meteorites toward the ocean. Some small percentage of this ice s never makes it that far; cruelly trapped by mountains or sub-ice barriers, it slowly evaporates and erodes away in the frigid climate, leaving only its bones, the detritus of space, on its surface. And on a few occasions, a few of the life forms of our planet have demonstrated their lack of intelligence by exploring this forbidding landscape. Coming across the members of this debris layer, they have sometimes recognized the bits and pieces for what they are; the loose bricks and mortar of our solar system, delivered to us while we wait. It is an amazing journey, rife with opportunities for loss and ruin. But in the end we have nothing short of a miracle: the very stuff of space delivered to our door, free of charge. We are indeed a very lucky species at times."(Harvey, 2003, 141) Dr. William Cassidy (1986) told me in the Antarctic that he thought that their would be over 100,000 meteorites recovered from the Antarctic and being that collection sites continue to be found and that there are now several countries with interest and activity in Antarctic meteorite recovery; 100,000 meteorites that might be recovered within the next century might be a conservative number.

Conclusion:

The meteorite recovery program is a priority because it is filling in the gap of getting samples from planets and extra-terrestrial bodies that have not yet had sample return missions. Currently the only samples with life, from Mars have been recovered from the Antarctic. Techniques of exploration that have been used in other locales such as cartography and map reading are used to select sites for exploration and reconnaissance and then after the meteorites are recovered their location is mapped out.(Mardon, 2004) Geography can be used to benefit the recovery of the precious meteorites from the Antarctic.

Currently Antarctic meteorites are giving thousands of samples from off of our planet. Cassidy says that, "so far, however, meteorites are the only actual samples of bodies other than the earth and moon that we can examine in the laboratory."(Cassidy, 2003, 129) Antarctic meteorites gave the first example of what has been theorized as evidence of life from a meteorite originally from Mars. By pairing up spectral signatures from meteorites as compared with parent bodies in the Inner Solar system it is possible to get physical geological samples that can be examined instead of costly extra-terrestrial sample return missions.

Bibliography:

(AMLAMP) www.lpi.usra.edu/research/amlamp/intro/intro.html

(Bayly and Stillwell, 1923) The Adelie Land meteorite. Australasian Antarctic Expedition 1911-1914, Science Reports Series, A, 4, 1-13.

(Cassidy, 1986) Cassidy, William A. Private communication, Lewis Cliff Ice Tongue, Antarctica, December 1986.

(Cassidy, 2003) Cassidy, William A. Meteorites, Ice and Antarctica. Cambridge University Press: United Kingdom, 2003.

(Folco, 2002) Folco, L., Capra, A. Chiappini, M., Frezzotti, M., Mellini, M. and Tabacco, I.E. The Frontier Mountain meteorite trap (Antarctica). Meteoritics and Planetary Science 37, 209-228.

(Harvey, 1998) Harvey, Ralph P; Dunbar, Nelia W; McIntosh, William C; Esser, Richard P; Nishiizumi, Kuni; Taylor, Susan; Caffee, Marc W. (1998) Meteoritic event recorded in Antarctic ice. Geology, vol.26, 607-610.

(Harvey, 2003) Harvey R. The Origin and Significance of Antarctic Meteorites. Chemie der Erde/Geochemistry, vol. 63, issue. 2, pp. 93-147.

(Kojima, 2001) Kojima, H; Kaiden, H; Yada, T "Meteorite search by JARE-39 in 1998-1999 season." Meteorite Research, 13, 1-8.

(Lucchitta, 1987) Lucchitta, BK;Bowell, J; Edwards, K; Ferguson, HM "Multispectral Landsat images of Antarctica." United States of America Geological Survey Bulletin 8755-531X: B 1696.

(Mardon, 1985) "Personal Communication." University of Lethbridge - Geography Department, March 1985, Lethbridge, Canada.

(Mardon, 1986) "Personal Communication." Lewis Cliff Ice Tongue Antarctica, 1986.

(Mardon, 1987a) "The Use of Low Level Aerial Photography to Detect Surficial Blue Ice Meteorites." The Explorers Journal. New York, United States of America. September 1987, page 41.

(Cassidy, et al-and Mardon, 1987) "The Meteorite Concentration at the Lewis Cliff. Antarctica, Ice Tongue." Meteoritics, The Journal of the Meteoritical Society. United States. Volume 22, Number 4, December 1987, page 353.

(Cassidy, et al-and Mardon, 1987b & c) "The Meteorite Concentration at the Lewis Cliff, Antarctic, Ice Tongue." Co-author. Conference Proceedings- 50th Annual Meeting of the Meteoritical Society. July, 1987, Newcastle Upon-Tyne, United Kingdom.

(Mardon, 1987d) "Antarctic Meteorite Surface Traverse 1986- 1987: A Field Report." Proceedings of the Association of American Geographers- 1987 Annual Meeting. April 24, 1987, Portland, Oregon, United States of America.

(Mardon, 1987e) "Meteorite". Meteor News. Published by the 'American Meteor Society', United States of America. Number 76, January 1987, page 3.

(Mardon, 1987f) "The Use of Low Level Remote Remote Sensing to Directly Detect Surficial Blue Ice Meteorites." Proceedings of the 19th Lunar and Planetary Science Conference. Clearlake, Texas, March 1987, 723.

(Mardon, 1987g) "Geographic Research in the Antarctic by the United States Research Program. The 20th South Dakota State University Geography Conference. South Dakota, United States of America. March 1987. Submitted in print only- in absentia.

(Mardon, 1988a) "Photography for Meteorite Recovery". Meteor News. Published by the 'American Meteor Society'. United States of America. Number 78, July 1988.

(Mardon, 1988b) "1986- 1987 Antarctic Meteorite Program." Meteor News. Published by the 'American Meteor Society'. United States of America. Number 80, January 1988, page 4.

(Mardon, 1988c) "The Deposition of Pleistocene Placer Deposits Located in North America." Meteoritics. The Journal of the Meteoritical Society. United States of America. Volume 30, Number 5, December 1988.

(Mardon, 1988d) "The Deposition of Pleistocene Meteorite Placer Deposits Located in North America." Proceedings of the 51st Annual meeting of the Meteoritical Society- July 1988. Arkansas, United States of America.

(Mardon, 1988e) "Antarctic Remote Sensing: Airborne, Satellite and Archival Sensing Systems." Proceedings of the American Association of the Geographers [AAG]- 1988 Annual Meeting. Phoenix, United States of America. April 1988, page 121.

(Mardon, 1989) "The Use of Archived Antarctic Aerial Photography." Proceedings of the Association of American Geographers [AAG]- 1989 Annual Meeting. Baltimore, Maryland, United States of America. April, 1989.

(Mardon, 1990) "Mining Potential of Asteroid Belt: Two economic powers already investigating." Canadian Mining Journal. Co-author with T. Kryzanowski. April, 1990, page 43.

(Mardon, 1991) "Aeolian Transport of Antarctic Meteorites immediately after Impact: Implications for Current Transportation Models." Papers Presented to the Sixteenth Symposium on Antarctic Meteorites- National Institute of Polar Research, Tokyo, Japan. July 5-7, page 187

(Mardon, 1992a) "Potential Meteorite Recovery Locales within Russian Antarctic Logistical Capability." Abstract Volume of 23rd Lunar and Planetary Science Conference. March 16- 20. Houston, Texas, United States of America.

(Mardon, 1992b) "Meteor Flux Rates on Non-Terrestrial Planetary Bodies: Impacts on Planetary Formation." Planetary Systems: Formation, Evolution and Detection, December 7- 10, 1992, Caltech, Pasadena, California, United States of America.

(Mardon, 1995) "A proposal for a series of well organized and properly equipped scientific expeditions to the Antarctic by the Canadian Scientific and Academic Community." Polar Times. United States of America. Fall- Winter 1995, pages 22 to 23.

(Mardon, 1996) "Meteorite Recovery." Science. United States of America. Volume 274, Number 5286, October 18, 1996, page 329.

(Mardon, 1997) "The Use of Russian Strategic Radar Systems to detect infalling meteorites for the purpose of recovering meteorites. Meteor News. Published by the 'American Meteor Society'. United States of America. Spring 1997, page 3.

(Mardon, 1998) "Personal Communication-Telephone." With Chinese Antarctic Program, Edmonton, Alberta, Canada.

(Mardon, 1998a) "Means of cooperation between Japanese Antarctic meteorite recovery program and potential Russian Antarctic meteorite recovery program." ESO Workshop on Minor Bodies in the Outer Solar System: Abstract Volume. Garching, Germany, November 2- 5, 1998.

(Mardon, 1998b) "Annual fluctuations of fern-snow cover in meteorite moraine." ESO Workshop on Minor Bodies in the Outer Solar System: Abstract Volume. Garching, Germany, November 2-5, 1998.

(Mardon, 2003) "Proposal for a French Antarctic meteorite recovery program." Annual Meteoritical Society Meeting, July, 2003, Muenster, Germany.

(Mardon, 2004) "The use of geographic remote sensing, mapping and aerial photography to aid the recovery of blue ice surficial meteorites in Antarctica." Pacific Journal of Science and Technology: An interdisciplinary Journal of Research, theories, and Observation. 2004, #2, United States of America.

(Mardon, 2004b) "The use of archived aerial photography to locate large meteorites that were on the Lewis Cliff Ice Tongue." Pacific Journal of Science and Technology: An interdisciplinary Journal of Research, theories, and Observation. 2004, #2, United States of America.

(Moser, 2002) Moser, D "Meteorite hunt successful; Further tourist based searches likely." Antarctic Non-Government Activities News 65/04 Australian Antarctic Division, Kingston, Tasmania 7050

(Nagata, 1982) Nagata, T. "Search for Antarctic Meteorites." In Antarctic Geoscience,page 1049-1058, University of Wisconsin Press, Madison, USA; 1982.

(Roscoe, 1986) "Personal Communication." 1986, Portola Valley, California, United States of America.

(Russell, 2002) Russell, S.; Zipfel, J; Grossman, J;Grady,M. The meteoritical Bulletin #86, July 2002. Meteoritics Planetary Science 37, (supplement) A157-A184

(Satterwhite, 2004) "Personal Communication." 2004, Johnson Space Centre, Houston, Texas, United States of America.

(Whillans, 1983) Whillans, I.; Cassidy, W. A. "To catch a falling star: Meteorites and old ice." Science, 1983.

Referenced articles published by author as research related to topic:

* "The use of geographic remote sensing, mapping and aerial photography to aid the recovery of blue ice surficial meteorites in Antarctica." Pacific Journal of Science and Technology: An interdisciplinary Journal of Research, theories, and Observation. 2004, #2, United States of America.

*"The use of archived aerial photography to locate large meteorites that were on the Lewis Cliff Ice Tongue." Pacific Journal of Science and Technology: An interdisciplinary Journal of Research, theories, and Observation. 2004, #2, United States of America.

* "Proposal for a French Antarctic meteorite recovery program." Annual Meteoritical Society Meeting, 2003, Germany.

*"The Use of Russian Strategic Radar Systems to detect infalling meteorites for the purpose of recovering meteorites. Meteor News. Spring 1997, page 3.

*"Meteorite Recovery." Science. Volume 274, Number 5286, October 18, 1996, page 329.

*"A proposal for a series of well organized and properly equipped scientific expeditions to the Antarctic by the Canadian Scientific and Academic Community." Polar Times. Fall- Winter 1995, pages 22- 23.

*"1986- 1987 Antarctic Meteorite Program." Meteor News. Number 80, January 1988, page 4.

*"The Use of Low Level Aerial Photography to Detect Surficial Blue Ice Meteorites." The Explorers Journal. September 1987, page 41.

*"Photography for Meteorite Recovery". Meteor News. Number 78, July 1988.

*"Meteorite". Meteor News. Published by the 'American Meteor Society', Number 76, January 1987, page3.

*"Means of cooperation between Japanese Antarctic meteorite recovery program and potential Russian Antarctic meteorite recovery program." ESO Workshop on Minor Bodies in the Outer Solar System: Abstract Volume. Garching, Germany, November 2- 5, 1998.

*"Annual fluctuations of fern-snow cover in meteorite moraine." ESO Workshop on Minor Bodies in the Outer Solar System: Abstract Volume. Garching, Germany, November 2-5, 1998.

*"Meteor Flux Rates on Non-Terrestrial Planetary Bodies: Impacts on Planetary Formation." Planetary Systems: Formation, Evolution and Detection, December 7- 10, 1992, Caltech, Pasadena, California.

*Co-author. "Potential Meteorite Recovery Locales Within Russian Antarctic Logistical Capability." Abstract Volume of 23rd Lunar and Planetary Science Conference. March 16- 20. Houston.

*"Aeolian Transport of Antarctic Meteorites immediately after Impact: Implications for Current Transportation Models." Papers Presented to the Sixteenth Symposium on Antarctic Meteorites- National Institute of Polar Research, Tokyo. July 5-7, page 187.

*"The Use of Archived Antarctic Aerial Photography." Proceedings of the Association of American Geographers [AAG]- 1989 Annual Meeting. Baltimore, Maryland, April, 1989.

*"The Deposition of Pleistocene Pacer Deposits Located in North America." Meteoritics. The Journal of the Meteoritical Society. Volume 30, Number 5, December 1988.

*"The Deposition of Pleistocene Meteorite Placer Deposits Located in North America." Proceedings of the 51st Annual meeting of the Meteoritical Society- July 1988. Arkansas.

*"Antarctic Remote Sensing: Airborne, Satellite and Archival Sensing Systems." Proceedings of the American

Association of the Geographers [AAG]- 1988 Annual Meeting. Phoenix, April 1988, page 121.

*"The Use of Low Level Remote Remote Sensing to Directly Detect Surficial Blue Ice Meteorites." Proceedings of the 19th Lunar and Planetary Science Conference. Clearlake, Texas, March 1987, 723.

*"Geographic Research in the Antarctic by the United States Research Program. The 20th South Dakota State University Geography Conference. March 1987. Submitted in print only.

*"The Meteorite Concentration at the Lewis Cliff. Antarctica, Ice Tongue." Meteoritics, The Journal of the Meteoritical Society. Co-author. Volume 22, Number 4, December 1987, page 353.

*"The Meteorite Concentration at the Lewis Cliff, Antarctic, Ice Tongue." Co-author. Conference Proceedings- 50th Annual Meeting of the Meteoritical Society. July, 1987, Newcastle Upon-Tyne, United Kingdom. Referenced in Astronomy and Astrophysics Abstracts.

*"Antarctic Meteorite Surface Traverse 1986- 1987: A Field Report." Proceedings of the Association of American Geographers- 1987 Annual Meeting. April 24, 1987, Portland, Oregon.

Biography of author:

Dr. Austin Albert Mardon, C.M. was born in 1962 in Edmonton, Alberta, Canada. He has degrees from University of Lethbridge, B.A., Geography; South Dakota State University, M.Sc., Geography; Texas A & M University, M.Ed., Education; Greenwich University (Australia), Ph.D., Geography. He has received four medals the first one was the Antarctic Service Medal from the United States of America National Science Foundation, the second one was the Queen Elizabeth II Golden Jubilee Medal from the Canadian government, the third one was the Alberta Centennial Medal from the Alberta government and the fourth one was the Order of Canada. In addition he was awarded the Governor Generals Caring Canadian Award which was personally presented to him by the Canadian Governor General. During the last 17 years he has received numerous awards for his humanitarian contributions in Canada. He has authored, co-authored, translated and edited over 50 books and authored over 200 scholarly communications. He authored a letter that was published in Science in 1996 and Nature in 1996. During the austral summer of 1986-1987 he was a field member of the United States of America ANSMET expedition to the Lewis Cliff Ice Tongue, Antarctica. He conducted experiments related to ground truth of meteorites being discernable from terrestrial rocks found lying on the surface of the ice. He has been presented to the Alberta Legislature, the Texas State Legislature and the Canadian federal Senate. In 1996 he was personally presented to Pope John Paul II in the Vatican. He is Corresponding Member of the Explorer's Club and a member of the International Academy of Astronautics. Currently he is the founding and current director of the Antarctic Institute of Canada. He serves on several volunteer boards in Edmonton where he currently resides.

www.ingramcontent.com/pod-product-compliance
Ingram Content Group UK Ltd.
Pitfield, Milton Keynes, MK11 3LW, UK
UKHW041141300926
15471UKWH00011B/87

9 781894 573153